Waves and Photons

Waves and Photons

AN INTRODUCTION TO QUANTUM OPTICS

EDWIN GOLDIN
Department of Physics
Bethany College
Bethany, West Virginia

John Wiley & Sons
New York / Chichester / Brisbane / Toronto / Singapore

Library of Congress Cataloging in Publication Data:

Goldin, Edwin, 1933-
Waves and photons.

(Wiley series in pure and applied optics, ISSN 0277-2493)
Includes index.
1. Quantum optics. I. Title. II. Series.

QC446.2.G64 1982 535′.15 82-10991
ISBN 0-471-08592-8

Printed in the United States of America

10 9 8 7 6 5 4 3 2 1

For
MARJORIE
a universal being

Preface

Written as a self-contained primer of modern optical theory, this book should appeal to undergraduate students of science and engineering (who have already taken courses in introductory physics and calculus) and to optical engineers, designers, and practitioners in industry and the professions. The latter part of the text should challenge graduate students studying quantum optics.

The book is not a compendium of all the multifaceted aspects of light, but rather a presentation of the theory by which these phenomena are understood. In this approach, the groundwork for the quantum properties of light begins with the classical picture. Ideas concerning wave packets, coherence, laser light, energy formulation, and light emission are introduced early and woven into the classical theory in preparation for their use in quantum theory.

The initial chapters offer a review of elementary concepts of classical physical optics. Reference to quantum ideas appear early in the text in order to provide an initial familiarity for the subsequently presented quantum theory. The material moves in a progression from fundamental physics to preparation for graduate study in quantum optics.

The design of the content is vertical. Application of physics to this area of optics builds across several layers of study, including mechanics and electricity and magnetism. Emphasis is on concepts and interpretation; mathematical procedures, such as Fourier series, are introduced only to enhance the physics. Finally, concepts about light are made as visual as possible through the use of many drawings and physical examples, especially for the new or more abstract ideas about coherence and light energy.

I have tried to make quantum theory accessible to the beginner. The fundamental elements of the Dirac formulation are slowly introduced and explained with free use of analogy. While this approach is more abstract than the Schrödinger wave theory, it is concise and mathematically simple—the only tools required are those of vector algebra.

Since the intention is to teach physics, the book begins with elementary plane waves and the classical wave theory of light is then carefully developed. The failure of this theory to account originally for the emerging photon provides the rationale for introducing quantum ideas.

Quantum theory is applied to the simple harmonic oscillator in order to establish minimum uncertainty (coherent) states. A similarity between oscillators and electromagnetic fields emerges by which the field can be quantized to display its photon character. There is a final indulgence in the application of these results to correlation functions and a description of laser and Gaussian light.

I hope this look at light will give the reader confidence to pursue an interest in the field and to brave the literature found in the journals.

No effort is a solo endeavor. Fond memories of my harshest critic, Dr. Judith Bregman, are recalled in writing this work. My gratitude goes to Dr. Donald B. Scarl, also of Polytechnic Institute of New York, who taught me to be coherent and precise in understanding quantum optics. I thank Dr. Hellmut J. Juretschke, former Head of the Department, who supported and kept faith in an old graduate student.

Most of all I am grateful to my family—my wife, Marjorie, and my children, Philip, Martin, Lauren, and Amanda—for contributing patience, love, and aid in the long hours of writing. To my friends, especially Dr. Thomas Heed and Mrs. Donna Peterson, special thanks for taking my thinking out loud.

Finally, I offer appreciation to Beatrice Shube for her warmth, guidance, and experience as an editor. I also wish to acknowledge the stability, tradition, and values of Bethany College where I have put the final touches to this book.

EDWIN GOLDIN

Bethany, West Virginia
July 1982

Contents

1

Anatomy of an Oscillation

If optics did not exist there would still be a wave theory. The wave function would still appear as the solution to a class of partial differential equations whose properties would be explored for their own sake. Application to mechanics alone would provide waves with meaning. Actually, their properties lend well to the physical characteristics of a large set of events in nature.

In optics, wave theory is a panacea. Except for Isaac Newton's attempt at a corpuscular explanation, we have seen about one hundred years of a wave theoretical basis for light capped by Clerk Maxwell's demonstration of the electromagnetic character of this radiant energy. Thus, it does not come lightly that wave theory fails where the photon succeeds. Still it is the wave, in a quantized formulation, that helps predict the photon and preserves the unity of both particle and wavelike features for light.

1.1 FUNCTIONS OF A SINUSOID

One of the basic features of a wave is its repetitiveness, a quality that is nicely demonstrated by a sinusoidal function. A plot of $y = \sin\theta$ in Fig. 1.1 represents the cyclic nature of $y(\theta)$ within any interval of θ and $\theta + 2\pi$. Mathematically, $y(\theta + 2\pi)$ returns to its former value

$$y(\theta) = \sin\theta = \sin(\theta + 2\pi) \tag{1.1}$$

There are many functions that satisfy this requirement; this just happens to be a simple and useful one. Also cyclic is

$$y(\theta) = \cos\theta = \cos(\theta + 2\pi) \tag{1.2}$$

which is a sine curve displaced backwards through one quarter of a cycle or, what is the same thing, a sine function advanced in phase by one quarter of

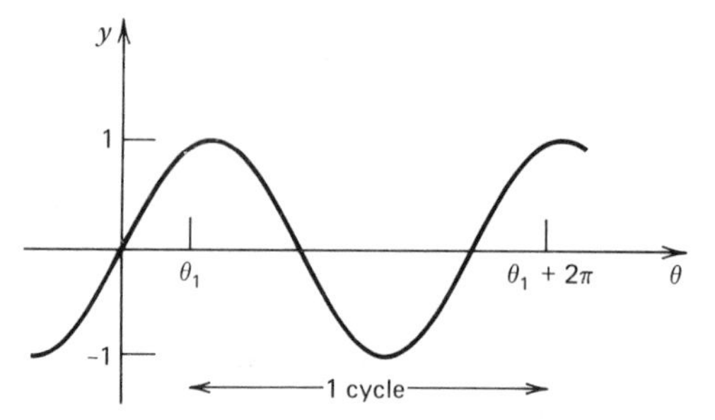

Figure 1.1. A graph of $y(\theta) = \sin\theta$.

a cycle. Figure 1.2 illustrates this for the functions

$$f = \cos\theta = \sin\left(\theta + \frac{\pi}{2}\right) = \sin\theta' \tag{1.3}$$

or

$$g = \sin\theta' = \cos\left(\theta' - \frac{\pi}{2}\right) = \cos\theta \tag{1.4}$$

where $\theta' = \theta + \pi/2$. The tangent function is periodic but is not very useful in describing waves since it is not only discontinuous but has an undefined value in its cycle. Any linear sum of sines or cosines and sines and cosines

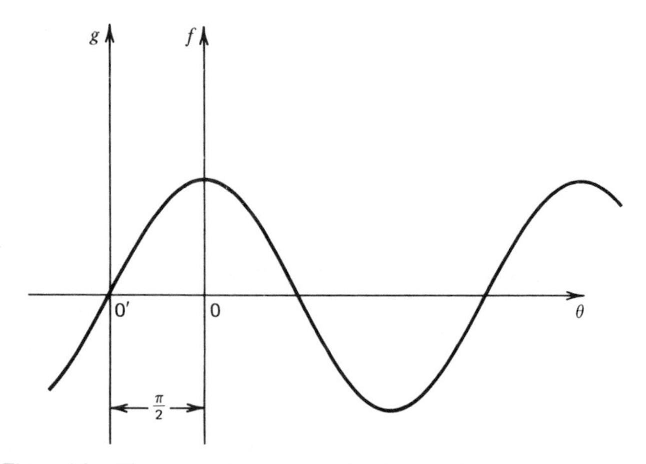

Figure 1.2. The sine and cosine function displaced by one-quarter cycle.

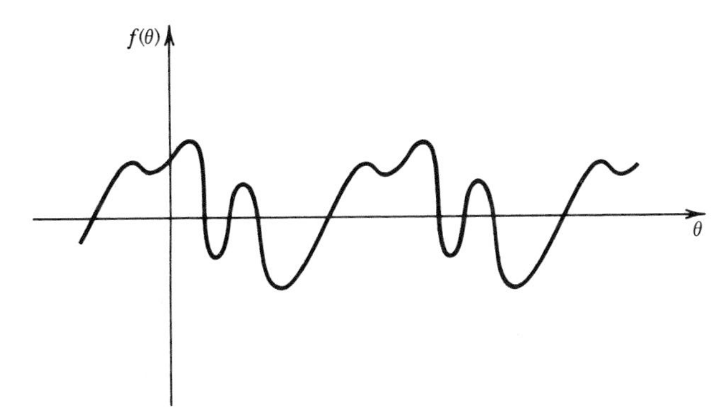

Figure 1.3. A more general periodic function $f(\theta)$.

will produce more general periodic functions, as we shall see later (Fig. 1.3). Among these functions will be $e^{i\theta}$, the exponential with an imaginary exponent, which will be quite useful in demonstrating the unfolding wave theory.

Whereas θ is an independent variable, the constant 2π or any other constant will be referred to as the constant phase or initial phase making $(\theta + \alpha)$ the total phase of the function. Hence, the more general sinusoidal function

$$y = \sin(\theta + \alpha) \tag{1.5}$$

where α is a constant, can be pictured as a sine function displaced backwards along the θ axis by the initial phase α, or as a sine function for which the origin has been shifted forward by α. Either change yields the same effect, as seen in Fig. 1.4, for

$$y = \sin(\theta + \alpha) \qquad y' = \sin\theta' \tag{1.6}$$

In the simplest sense, the phase constant is dependent upon where the origin is placed. The variable phase θ will be found in later applications to be a function of time, or space, or both time and space coordinates, $\theta(x, t)$. In a comparison of phase between two functions $y_1 = f(\theta_1)$ and $y_2 = f(\theta_2)$, the quantity, $\delta = \theta_2 - \theta_1$ is regarded as the phase difference.

Since the sine (or cosine) function varies in value between ± 1, the coefficient A takes on the meaning of maximum value of the function. Hence, we have the function

$$y = A\sin(\theta + \alpha) \tag{1.7}$$

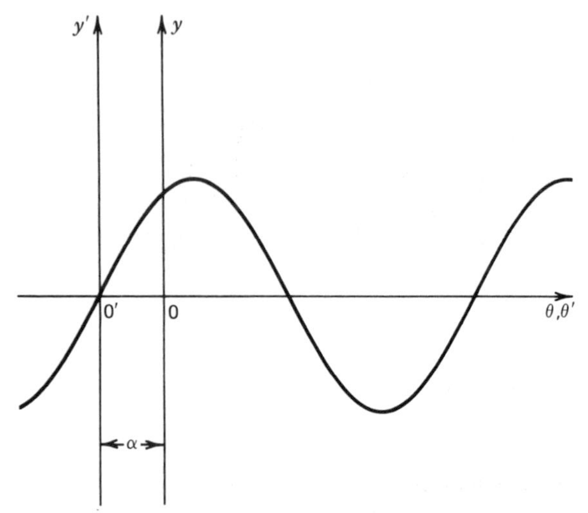

Figure 1.4. The initial phase as equivalent to a shift in the origin.

oscillating in Fig. 1.5 between upper and lower values fixed by $\pm |A|$. Generally the amplitude is a positive number, but it may be necessary to deal with amplitudes that are also a function of time and with amplitudes that are complex numbers. This simple constant amplitude A will in large measure support the connection of classical wave theory in its relation to energy. It will also play an important role in the limitations of the theory and take on a curious relation to photons in quantum optics.

1.2 PERIODICITY

Plotting the sinusoidal function (1.7), as in Fig. 1.6, the cyclic character has a length $\theta_2 - \theta_1 = 2\pi$, where θ_1 and θ_2 are the two closest consecutive values that represent the same cyclic point in the function $y(\theta)$. This is simply the original definition of periodicity, namely, $y(\theta_1) = y(\theta_2)$.

While it is tempting to stop at an intermediate point θ_3, since it is also true that $y(\theta_1) = y(\theta_3)$, these points are not a full cycle apart in periodicity as can be seen from the first derivative.

$$\left(\frac{dy}{d\theta}\right)_{\theta=\theta_1} \neq \left(\frac{dy}{d\theta}\right)_{\theta=\theta_3} \tag{1.8}$$

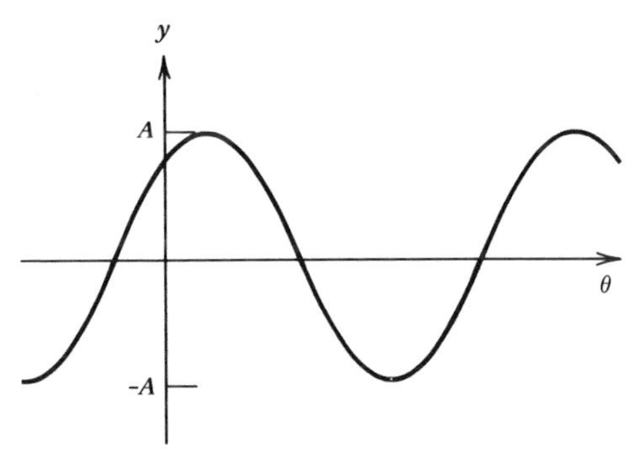

Figure 1.5. Extreme values of $y(\theta)$.

Therefore, a full cycle is achieved not only when

$$y(\theta_1) = y(\theta_1 + \Delta\theta) \tag{1.9}$$

but also when

$$\left(\frac{dy}{d\theta}\right)_{\theta_1} = \left(\frac{dy}{d\theta}\right)_{\theta_1 + \Delta\theta}$$

where $\Delta\theta$ is the smallest nonzero value for which this is true. More commonly stated, a full cycle is traversed when the function returns to its initial value and is about to change in the same way.

If θ is a function of x, as will be encountered in applications to physical systems, then in a graph of

$$y = A \sin[\theta(x) + \alpha] \tag{1.10}$$

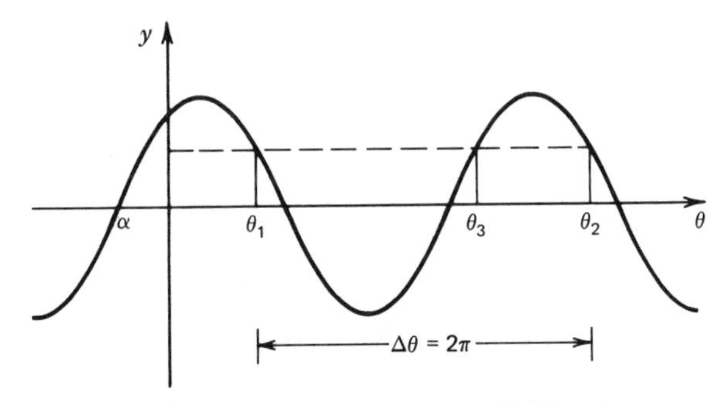

Figure 1.6. Cyclic length of a sinusoidal function.

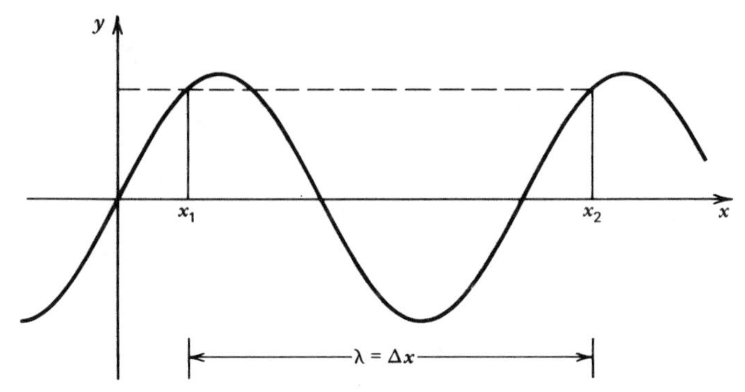

Figure 1.7. Wavelength of a cyclic function.

against an x axis (Fig. 1.7), the length of a cycle is rendered by λ where $x_2 - x_1 = \lambda$ and $\theta(x_2) - \theta(x_1) = 2\pi$. In particular, λ will stand for wavelength when these functions are employed to describe waves.

By analogy with the preceding description, if instead of being spatially dependent θ is a function of time, then the time length in Fig. 1.8 of the full cycle for

$$y = A\sin[\theta(t) + \alpha] \tag{1.11}$$

is T, the period, where $\Delta t = t_2 - t_1 = T$ and $\theta(t_2) - \theta(t_1) = 2\pi$.

The wavelength and period each represent a defined quantity per cycle. Their reciprocals describe the number of cycles contained per unit of their respective quantities.

Thus, $1/T$ is the number of cycles/s (Hertz) which is dubbed frequency $f = 1/T$ and $1/\lambda$ specifies the number of cycles/unit length.

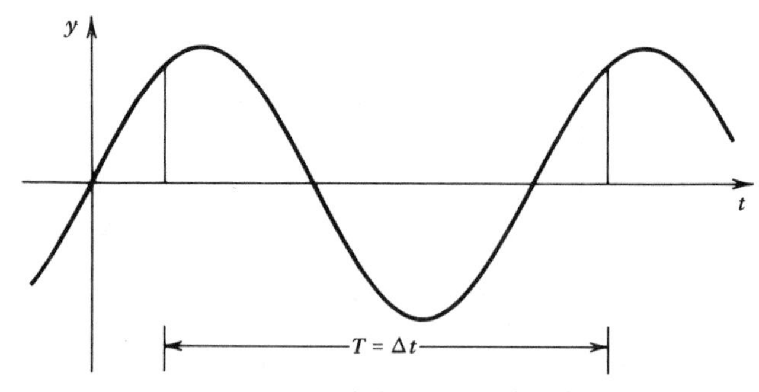

Figure 1.8. Period of a cyclic function.

Since both of these quantities are related to the angular term in the cyclic function, their measure in radians rather than cycles is particularly convenient. Therefore, we define,

$$\omega = \frac{2\pi}{T} \tag{1.12}$$

as the *angular frequency* (or more commonly the angular speed) representing the number of radians/second. In quantum theory this seemingly innocent quantity will take on an intimate relationship with energy.

Analogously, the term *wave number*

$$k = \frac{2\pi}{\lambda} \tag{1.13}$$

is invented to specify the number of radians/unit length. Later this wave number will become a vector describing the propagation direction of waves, and will be connected with the momentum characteristics of both classical waves and quantum particles.

1.3 A SUMMARY OF SINUSOIDS

On a most elementary level trigonometric quantities are introduced as defined ratios of the sides of a right triangle. In order to generalize these definitions a unit circle is invoked whereby, for instance, the sine of any angle can be defined as the y displacement of a point on a circle made by the position of a unit radius for that angle, as shown in Fig. 1.9. By plotting the projection on the y axis of each point on the circle corresponding to θ, we generate again a "picture" of the sine function. This rendition will be handy in solving the differential equation for a harmonic oscillator.

A word about the motion of this cyclic sinusoidal function is in order. We have looked at the elementary characteristics of these static sinusoidal curves, but did not refer to them as waves since nothing was waving, that is, in motion. Actually they are waves—mathematical waves having zero velocity. It is as if the observer and wave both have the same velocity, making the relative velocity between them zero.

As a mathematical entity there are no restrictions on the overall length of the wave; it could be infinitely long or merely an infinitesimal segment and still be describable by the appropriate sinusoidal function in that range. In using mathematics to describe periodic physical events, however, we shall find the length or duration of a wave to be quite significant.

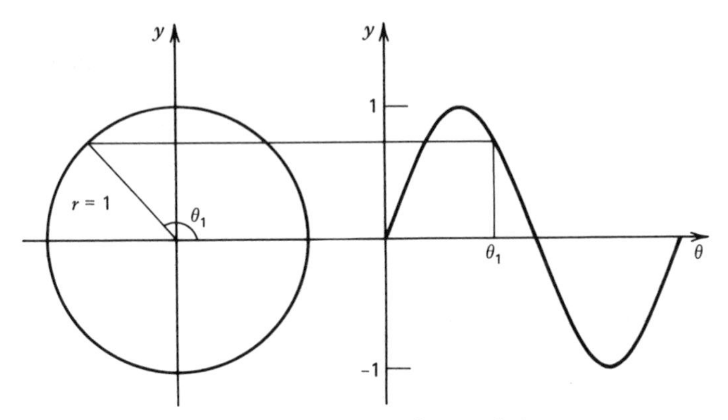

Figure 1.9. The unit reference circle.

There may be some physical occurrence that behaves in such a fashion that the variation of some property obeys

$$y(\theta) = A\sin(\theta + \alpha) \tag{1.14}$$

But if the match is only close we will be prepared to improve the description at every turn and even replace the mathematical model if necessary. A history of classical optics could be written as an echo of this statement reverberating on into the quantum model of light.

To this extent we look upon wave theory as a model. The waves it predicts are mathematical constructs of varying amplitudes $y(\theta)$, the collection of which form an envelope having properties that must be tested by experiment. The physical understanding and description are the goal, not the rationalization that the mathematical model is the reality itself. With this caution we proceed to an idealized physical system—the harmonic oscillator.

1.4 SIMPLE HARMONIC OSCILLATORS

Among cyclic motion one of the simplest motions is straight-line repeating oscillations for which the acceleration of the object is proportional to its displacement from equilibrium and oppositely directed.

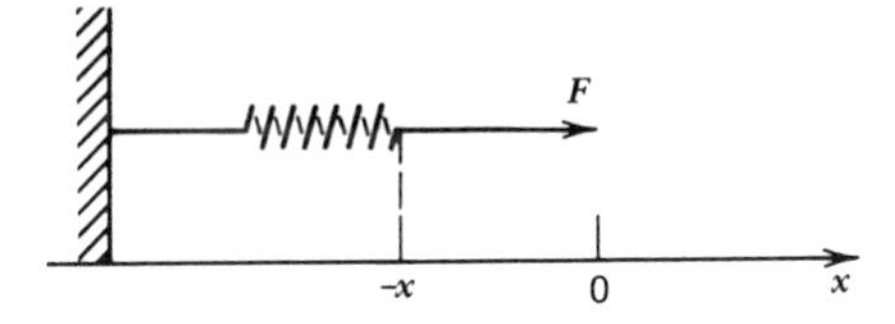

Figure 1.10. Force exerted by a compressed spring.

This condition

$$a = \frac{d^2x}{dt^2} \sim -x \tag{1.15}$$

is closely approximated by the motion of a mass on the end of a real spring. If a spring, extended or compressed from its natural length, provides a force linear with its displacement and directed towards its undisturbed position, it satisfies Hooke's law

$$F = -Kx \tag{1.16}$$

where K is the constant of proportionality in the linear relation between F and $-x$. The spring constant K is always positive since a positive displacement x produces a negative restoring force while a negative displacement, as shown in Fig. 1.10, provides a positive one.

For simplicity we arrange a mass on the end of a horizontal spring (Fig. 1.11), with the other end tied to an immobile wall. The mass rests on a frictionless horizontal surface and is displaced a distance A from the undisturbed natural length of the spring. Under these conditions the ensuing motion will be referred to as simple harmonic motion (SHM).

Upon release, the mass will oscillate back and forth attaining a position $x = -A$ halfway through one cycle of motion and $x = A$ at the completion

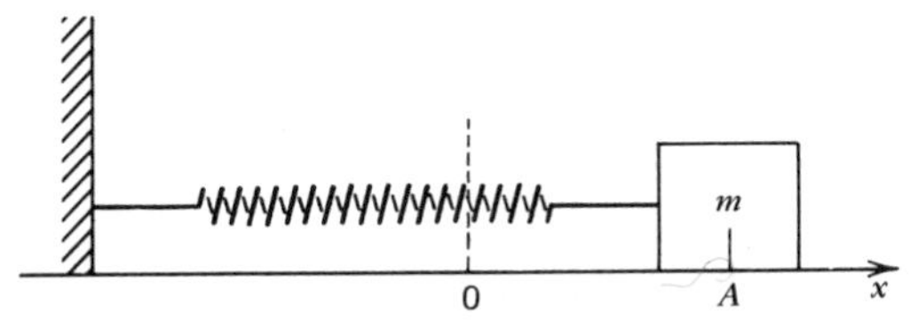

Figure 1.11. A mass on the end of a stretched spring.

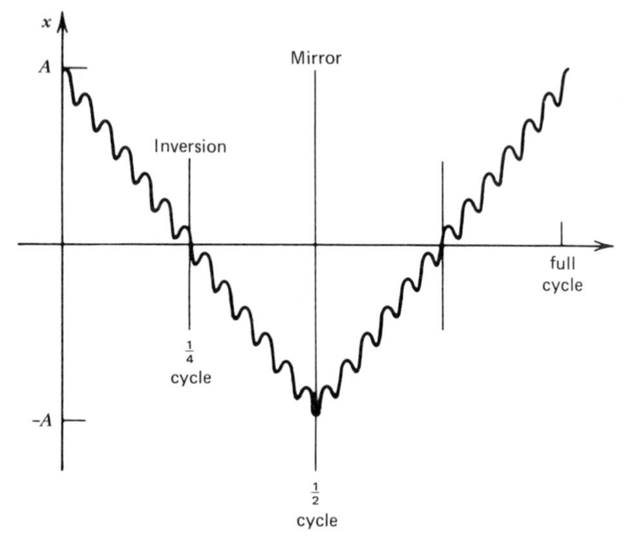

Figure 1.12. Symmetry diagram for displacement in SHM.

of that cycle. The dynamics of the system can be understood by looking at the physics of just one quarter of the cycle from $x = A$ to $x = 0$.

1. A clock starts reading time from $t = 0$, the instant we let the mass go from $x = A$.
2. In this position $x = A$, the restoring force provided by the spring is a maximum magnitude $F_{\max} = -KA$ in the negative direction.
3. This force promotes an acceleration which is also negative, that is, directed toward $x = 0$.
4. The magnitude of both the force and the acceleration diminishes as the mass reduces its displacement.
5. At the center the displacement is zero; the force and acceleration have likewise become zero. The mass has moved through one quarter of its cycle.

Given the symmetry of the setup, namely, for negative displacements the force and acceleration become positive, the second quarter of the cycle is the same as the first quarter but played in reverse with a minus sign (an inversion). At precisely one-half of the cycle $x = -A$, the force and acceleration have achieved a maximum value again, but this time in the positive direction, that is, toward the center. The symmetry should now predict the remaining half-cycle to be the same as the first half-cycle but

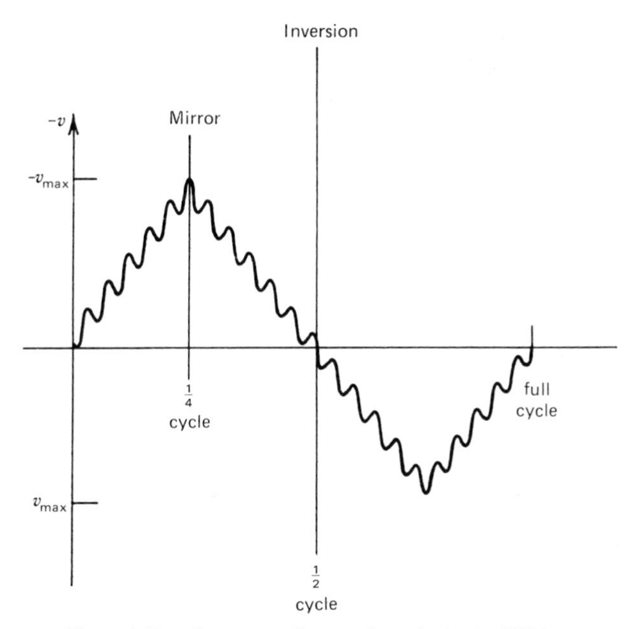

Figure 1.13. Symmetry diagram for velocity in SHM.

played in reverse (a mirror image). A diagram of these symmetries for the displacement, for example, might look something like Fig. 1.12. The velocity which we neglected (Fig. 1.12) is slightly different in its sequence of symmetry in that it begins at $x = A$ with a zero value, and reaches a maximum negative value at the end of one-quarter cycle; then in mirror symmetry drops back to zero at $x = -A$, one-half cycle. From there it inverts its direction replaying its magnitude back through the second half-cycle (an inversion). A diagram of the velocity appears in Fig. 1.13. While the symmetries rather nicely display the cyclic nature of the displacement, force, acceleration, and velocity for SHM and even hint at the

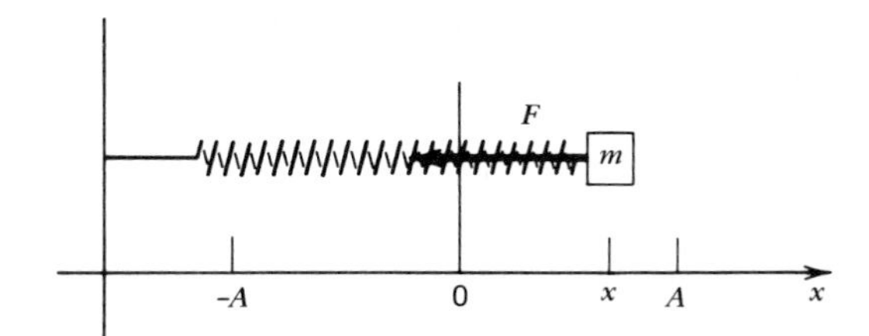

Figure 1.14. A mass in SHM at a general position **x**.

possibility of sinusoidal descriptions, a little mathematical rigor and further physical analysis will be helpful to firmly establish the resulting description.

The mass in Fig. 1.14 is acted upon by a net force having a Hooke's law nature $F = -Kx$ so that the acceleration of the mass from Newton's second law is

$$a = \frac{d^2x}{dt^2} = \frac{F}{m} = -\frac{K}{m}x \tag{1.17}$$

Since K and m are always positive constants, the factor K/m can be replaced by a new positive constant, say ω^2, such that

$$\omega^2 = \frac{K}{m} \tag{1.18}$$

Inserting this into Eq. (1.17) and rearranging leaves us with a second-order linear differential equation with constant coefficients

$$\frac{d^2x}{dt^2} + \omega^2 x = 0 \tag{1.19}$$

whose solution should describe everything about our ideal classical oscillator. Just to remove any doubt about why we call it ideal, we need only remember there is no external force, no mass to the spring, no resistance of any kind to the motion of the mass and the spring, and the wall is infinitely massive so that the system does not become a two-body problem. Each of these features and more can be brought into the description. For our purposes, the relation of SHM to waves and photons, we do not need to deal with these complexities.

1.5 SOLUTIONS FOR SHM

Knowing some theory of differential equations, we can write immediately the solutions for Eq. (1.19); knowing a little less we could integrate the equation twice (which is a bit tedious). As a first attempt we could simply guess at the solution.

We are looking for a function $x(t)$ which when differentiated twice gives back the same function with a minus sign. Since we are already disposed toward sinusoidal functions, an investigation of their derivatives reveals

$$\frac{d(\sin\theta)}{d\theta} = \cos\theta \tag{1.20}$$

and

$$\frac{d(\cos\theta)}{d\theta} = -\sin\theta$$

We have already seen in Eqs. (1.3) and (1.4) that the sine and cosine are similar but for a phase shift of $\pi/2$; the derivative serves the purpose of adding just such a phase shift to the function. Advancing another $\pi/2$ phase shift with the second differentiation, the slope of the slope gives a total shift of π from the original functions, which introduces a minus sign. The second derivative therefore is

$$\frac{d^2\sin\theta}{d\theta^2} = -\sin\theta$$

$$\frac{d^2\cos\theta}{d\theta^2} = -\cos\theta \tag{1.21}$$

This is just the property that we are seeking to satisfy the differential equation (1.19)! By assuming solutions in the form of

$$x(t) = C\begin{cases}\sin\theta \\ \text{or} \\ \cos\theta\end{cases} \tag{1.22}$$

where θ is a function of time and C is an arbitrary constant, we wish to show that one or the other of Eq. (1.22) satisfies the differential equation (1.19). If we choose

$$x(t) = C\cos\theta(t) \tag{1.23}$$

the first derivative of this function with respect to time (which is the velocity) becomes

$$\frac{dx}{dt} = \frac{dx}{d\theta}\frac{d\theta}{dt} = (-C\sin\theta)\frac{d\theta}{dt} \tag{1.24}$$

As long as we impose upon θ to be linear with time, or what is the same thing, that $d\theta/dt$ is a constant, the second time derivative (or acceleration) is

$$\frac{d^2x}{dt^2} = \frac{d^2x}{d\theta^2}\left(\frac{d\theta}{dt}\right)^2 = (-C\cos\theta)\left(\frac{d\theta}{dt}\right)^2 \tag{1.25}$$

Inserting the expression Eq. (1.25) for the acceleration and Eq. (1.23) for the displacement in the differential equation (1.19) we obtain

$$-C\cos\theta\left(\frac{d\theta}{dt}\right)^2 + \omega^2 C\cos\theta = 0 \tag{1.26}$$

which requires that

$$\left(\frac{d\theta}{dt}\right)^2 = \omega^2 \tag{1.27}$$

for the Eq. (1.26) to be identically satisfied for all angles θ. This leaves us with

$$\omega = \pm\frac{d\theta}{dt} \tag{1.28}$$

which upon integration yields

$$\theta = \pm\omega t + \alpha \tag{1.29}$$

Substituting this result (1.29) for the phase angle in the assumed solution (1.23) yields

$$x(t) = C\cos(\pm\omega t + \alpha) \tag{1.30}$$

The negative sign which appears for $d\theta/dt$ in Eq. (1.28) has the significance of describing the cycle for SHM running backwards. Since the simple symmetry of this cycle is such that it is physically the same cycle forwards as backwards we shall remove the negative sign as its presence here is unnecessary, and use simply

$$x(t) = C\cos(\omega t + \alpha) \tag{1.31}$$

to represent the solution for SHM. (In the analysis of wave motion, we will encounter a similar duality in the signs for the change of phase of a traveling wave in which case it cannot be ignored but rather will lead to two waves traveling in opposite directions.)

The solution for the displacement (1.31) is not only the most general solution to the differential equation (1.19) but is also identical to the other possible choice, $C\sin\theta(t)$. If we express the arbitrary phase constant α as, $\alpha = \alpha' - \pi/2$, the solution (1.31) becomes

$$x(t) = C\cos(\omega t + \alpha' - \pi/2) = C\sin(\omega t + \alpha') \tag{1.32}$$

making either form the same solution.

That the solution is the most general solution possible is guaranteed because it has two arbitrary constants with which we can describe the oscillator starting in any position with any velocity. We shall see this in a more fundamental way when we discuss the principle of superposition. For now, let us summarize the properties of the solution.

We have assumed a general solution to Eq. (1.19) with two arbitrary constants C and α to be determined by the conditions of the physical problem. In the process of obtaining these results we have established that

$$\omega = \frac{d\theta}{dt} = \sqrt{\frac{K}{m}} \tag{1.33}$$

showing ω to be an angular frequency defined by the constant rate of change of the phase. Since the phase must change by 2π in the time it takes to complete one cycle

$$\omega = \frac{2\pi}{T} = 2\pi f \tag{1.34}$$

where T is the period and f the frequency as defined in Section 1.1. It is interesting that ω, after being innocently introduced as a constant of the system (in this case, $\omega^2 = K/m$), takes on the meaning of angular frequency describing the oscillatory motion. Being defined in this manner as an angular speed specifying the rate of change of the phase, it need not be confused with angular velocity, which is a vector describing rotational motion.

In the general solution the displacement of the oscillator is again written as

$$x(t) = C\sin(\omega t + \alpha) \tag{1.35}$$

The velocity can be gained from this expression by taking the first derivative

$$v(t) = \frac{dx(t)}{dt} = \omega C\cos(\omega t + \alpha) \tag{1.36}$$

and likewise the acceleration through another differentiation becomes

$$a(t) = \frac{dv(t)}{dt} = \frac{d^2x(t)}{dt^2} = -\omega^2 C\sin(\omega t + \alpha) \tag{1.37}$$

which with the aid of Eq. (1.35) is

$$\frac{d^2x(t)}{dt^2} = -\omega^2 x(t) \tag{1.38}$$

not surprisingly the original differential equation. It now remains to develop a scheme for determining the arbitrary constants.

In Fig. 1.11 the mass was released from the position $x(0) = A$ with a velocity $v(0) = 0$ at the time $t = 0$. Using Eqs. (1.35) and (1.36) the initial conditions require

$$x(0) = C \sin \alpha = A$$

$$v(0) = \omega C \cos \alpha = 0 \tag{1.39}$$

making $\alpha = \pi/2$ and $C = A$, whereupon the particular solution for the oscillator starting in the position of maximum displacement is

$$x(t) = A \sin(\omega t + \pi/2) = A \cos \omega t \tag{1.40}$$

with the velocity

$$v(t) = -A\omega \sin \omega t = -v_{\text{max}} \sin \omega t \tag{1.41}$$

and the acceleration

$$a(t) = -A\omega^2 \cos \omega t = a_{\text{max}} \cos \omega t \tag{1.42}$$

The magnitude of the velocity is a maximum, $v_{\text{max}} = A\omega$, at $t = T/4$, or any odd multiple thereof. The magnitude of the acceleration is a maximum, $a_{\text{max}} = A\omega^2$, at any multiple of a half-cycle which is the position of extreme displacement. This is nicely summarized in Fig. 1.15.

If we wish instead to describe this motion for the oscillator starting at $x(0) = 0$ with $v(0) = A\omega$, the conditions would impose on Eqs. (1.35) and (1.36) that $\alpha = 0$ and $C = A$ giving the displacement function the form

$$x(t) = A \sin \omega t \tag{1.43}$$

In general, therefore, for any oscillator in SHM, given the initial position and velocity at the start of the clock ($t = 0$) to be

$$x(0) = C \sin \alpha \quad \text{and} \quad v(0) = \omega C \cos \alpha \tag{1.44}$$

we obtain

$$\alpha = \tan^{-1} \frac{\omega x(0)}{v(0)} \tag{1.45}$$

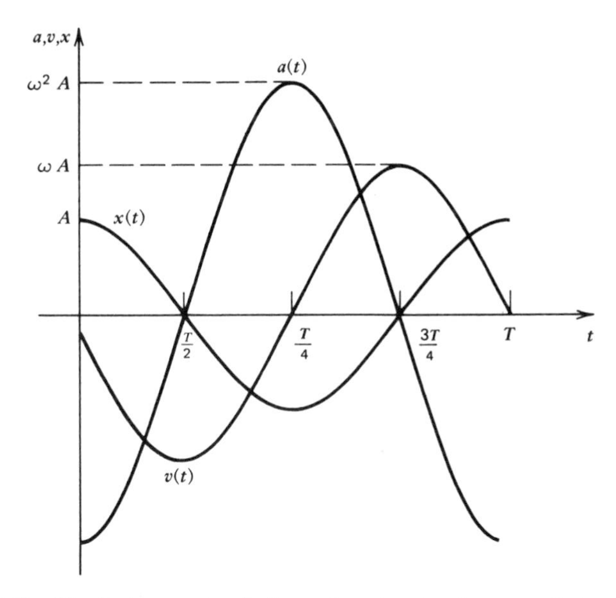

Figure 1.15. The displacement, velocity, and acceleration for an oscillation in SHM.

and

$$C = \sqrt{[x(0)]^2 + \left[\frac{v(0)}{\omega}\right]^2} \tag{1.46}$$

for the complete solution of the elementary oscillator described by

$$x = C\sin(\omega t + \alpha) \tag{1.47}$$

1.6 SUPERPOSITION

Being linear the differential equation (1.19) for SHM will admit solutions which can be linearly summed and still be solutions of the equation. Since either $\sin \omega t$ or $\cos \omega t$ individually satisfies the differential equation, their linear combination

$$x = A \sin \omega t + B \cos \omega t \tag{1.48}$$

where A and B are arbitrary constants, also represent a solution. With the aid of the reference triangle in Fig. 1.16, a redefinition of the arbitrary

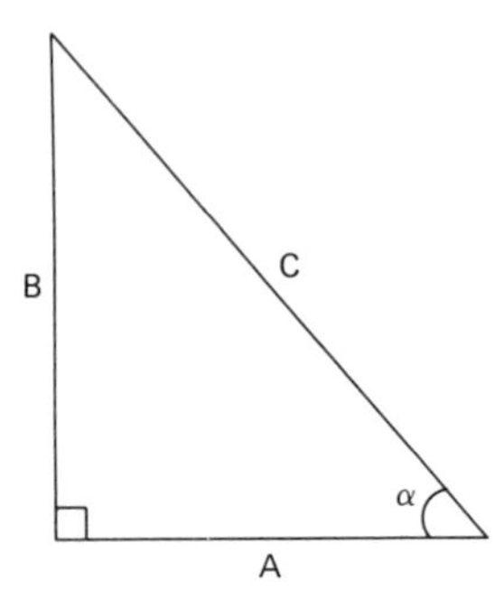

Figure 1.16. Right-angle reference triangle.

constants as $A = C\cos\alpha$ and $B = C\sin\alpha$ turns the above superposition of functions (1.48) into the now familiar general solution

$$x = C\sin(\omega t + \alpha) \tag{1.49}$$

Any further superposition of additional solutions of the sinusoidal form having the same angular frequency will produce nothing new for SHM, but rather recompose into either of the above general solutions. For example, if

$$x = \sum_i (a_i \sin\omega t + b_i \cos\omega t) \tag{1.50}$$

then

$$x = \sum_i c_i \cos(\omega t + \alpha_i) = C\cos(\omega t + \alpha)$$

where

$$C = \sqrt{\left(\sum_i a_i\right)^2 + \left(\sum_i b_i\right)^2}$$

and

$$\alpha = \tan^{-1}\frac{\sum_i b_i}{\sum_i a_i}$$

In this case, (1.50), the superimposed functions are linearly dependent in that it is possible to have

$$\sum_i c_i \cos(\omega t + \alpha_i) = 0 \tag{1.51}$$

for any t, where not all the c_i are zero; whereas it is not possible to achieve this with

$$x = A \sin \omega t + B \cos \omega t \tag{1.52}$$

making the $\sin \omega t$ and $\cos \omega t$ independent solutions. The notion of independent solutions will take on a large importance in building general solutions to the wave equation via Fourier series and Fourier integrals. Further, in quantum theory the superposition of linearly independent states will allow us to define a multidimensional vector space in which we will compose general states of a system. In particular, quantum states for the harmonic oscillator will be superimposed to establish a coherent state which can be applied to quantized electromagnetic fields.

1.7 SIMPLE HARMONIC ENERGY

To extend a spring requires work. The work performed on a spring undergoing a displacement from x through an infinitesimal change to $x + dx$ during which time an external Hooke's law force $F = Kx$ is acting (Fig. 1.17) is

$$dW = F\,dx = Kx\,dx \tag{1.53}$$

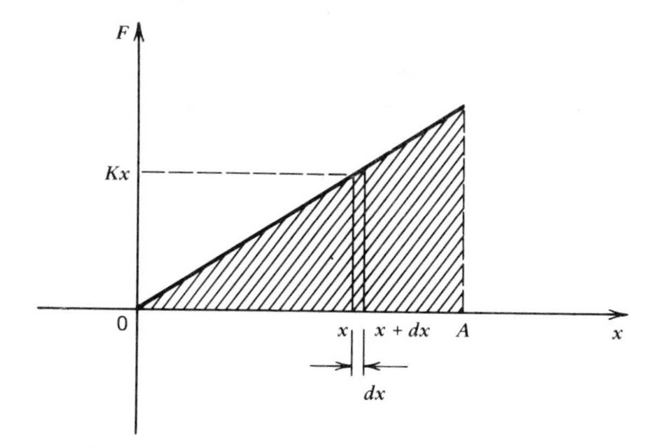

Figure 1.17. Work done under a Hooke's law force.

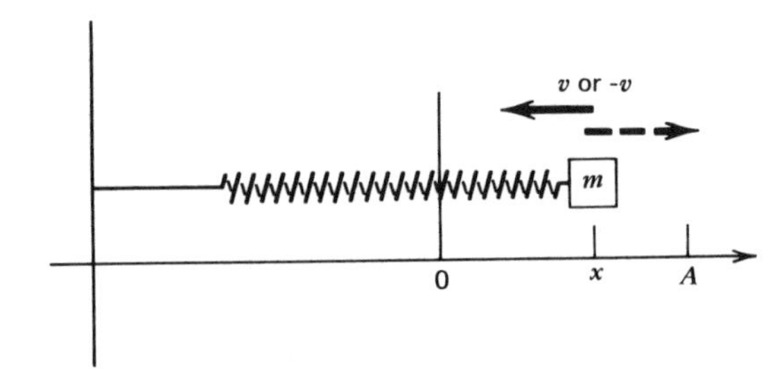

Figure 1.18. A mass with a general displacement and velocity in SHM.

If the spring is displaced from its equilibrium position $x = 0$ to any x, the work done becomes

$$W = \int_0^x Kx\,dx = \frac{Kx^2}{2} \tag{1.54}$$

This work (for an ideal frictionless spring) can be viewed as potential energy stored in the system. In the case of an oscillator displaced to a maximum $x_{\text{max}} = A$ and released from rest, the initial potential energy $\frac{1}{2}KA^2$ which is also the maximum potential energy, represents the total energy of the system. As the oscillator swings to any displacement x at which it may have a velocity v (Fig. 1.18), the total energy of the system by conservation of energy is composed of kinetic and potential energy

$$E_{\text{total}} = \tfrac{1}{2}Kx^2 + \tfrac{1}{2}mv^2 = \tfrac{1}{2}KA^2 = \text{const.} \tag{1.55}$$

where the total energy under ideal conditions remains constant. Passing through the equilibrium position $x = 0$ where the velocity is a maximum, the energy is all kinetic. The potential energy at that instant is zero. This energy picture is tidily displayed in Fig. 1.19 which shows a parabolic potential energy function in addition with another parabolic kinetic energy function producing a constant total energy. Since conservation of energy demands

$$\tfrac{1}{2}mv_{\text{max}}^2 = \tfrac{1}{2}KA^2 \tag{1.56}$$

it is evident using Eq. (1.33) that

$$v_{\text{max}} = \left(\frac{K}{m}\right)^{1/2} A = \omega A \tag{1.57}$$

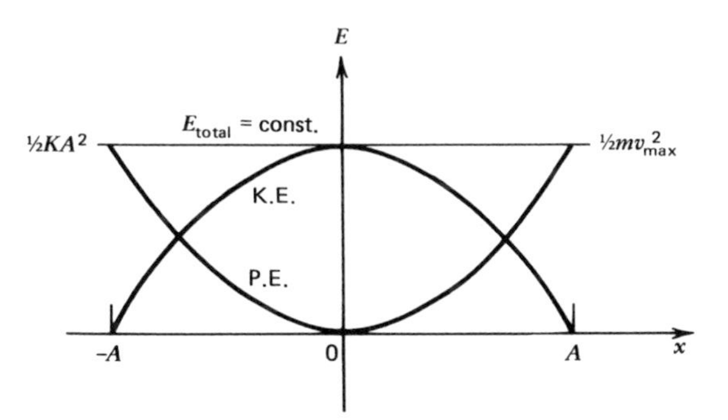

Figure 1.19. Parabolic energy functions for SHM.

Rewriting Eq. (1.55) as

$$v^2 + (\omega x)^2 = v_{max}^2 = \omega^2 A^2 \tag{1.58}$$

we have the equation of a circle of radius v_{max}. With the angle $\theta = \omega t + \alpha$ in Fig. 1.20, the projections on the axes are

$$\omega x = v_{max} \sin(\omega t + \alpha) = \omega A \sin(\omega t + \alpha) \tag{1.59}$$

and

$$v = v_{max} \cos(\omega t + \alpha)$$

which are the familiar solutions for SHM.

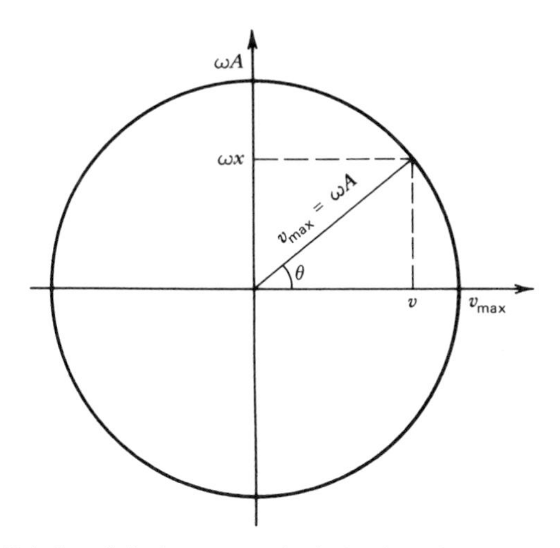

Figure 1.20. Relation of displacement and velocity through energy conservation.

More formally Eq. (1.58) can be solved directly by noting

$$v = \frac{dx}{dt} = \pm\omega\sqrt{A^2 - x^2} \tag{1.60}$$

or

$$\omega\, dt = \pm\frac{dx}{\sqrt{A^2 - x^2}}$$

which upon integration yields again

$$x = A\sin(\omega t + \alpha) \tag{1.61}$$

where $x = A\sin\alpha$ at $t = 0$.

An interesting formulation for the energy of a simple harmonic oscillator is obtained by introducing the momentum $p = mv$ in terms of which Eq. (1.55) becomes

$$E = \frac{m\omega^2 x^2}{2} + \frac{p^2}{2m} \tag{1.62}$$

In the case of an oscillator with unit mass ($m = 1$), the symmetry in the energy expression allows

$$E = \frac{\omega^2 x^2 + p^2}{2} = \tfrac{1}{2}(\omega x + ip)(\omega x - ip) \tag{1.63}$$

where $i = \sqrt{-1}$. The complex factors can be redefined by a new variable

$$a = (2\hbar\omega)^{-1/2}(\omega x + ip) \tag{1.64}$$

and its complex conjugate

$$a^* = (2\hbar\omega)^{-1/2}(\omega x - ip)$$

where $\hbar$ is a constant. Expressed in terms of these complex variables the energy succinctly is

$$E = \hbar\omega a^* a \tag{1.65}$$

which, although of little value here, in the classical analysis of SHM, it does show the classical energy to be related to the square of a variable. This formulation, however, takes on a useful role in the quantum analysis of the oscillator where the constant $h = 2\pi\hbar$ is Planck's constant and the complex amplitudes a and a^* are quantum operators casting the energy in the role of an operator.

The time dependence of the kinetic and potential energies are

$$\text{K.E.} = \frac{m}{2}\left(\frac{dx}{dt}\right)^2 = \frac{m\omega^2 A^2}{2}\cos^2(\omega t + \alpha) \tag{1.66a}$$

and

$$\text{P.E.} = \frac{K}{2}x^2 = \frac{KA^2}{2}\sin^2(\omega t + \alpha) \tag{1.66b}$$

A time average of the kinetic energy over one cycle appears as

$$\langle\text{K.E.}\rangle = \frac{\dfrac{m\omega^2 A^2}{2}\displaystyle\int_0^T \cos^2(\omega t + \alpha)\,dt}{\displaystyle\int_0^T dt} \tag{1.67}$$

From the identity $\cos^2\theta = \frac{1}{2} + (\cos 2\theta)/2$, where in this case $\theta = \omega t + \alpha$, the integral gives a contribution only for the $\frac{1}{2}$ term since the integral over any number of complete cycles of $\cos 2\theta$ must average to zero. This is seen in Fig. 1.21 where the $\cos^2\theta$ is shown as a plot of $\cos 2\theta$ displaced upward on the vertical axis by the factor one-half. In one cycle, $\theta = 2\pi$, the $\cos^2\theta$ completes two cycles over which the average (looking at the shaded area) becomes $\frac{1}{2}$. Thus the average kinetic energy is

$$\langle\text{K.E.}\rangle = \left(\frac{m\omega^2 A^2}{2}\right)\left(\frac{T/2}{T}\right) = \frac{\text{K.E.}_{\text{max}}}{2} \tag{1.68}$$

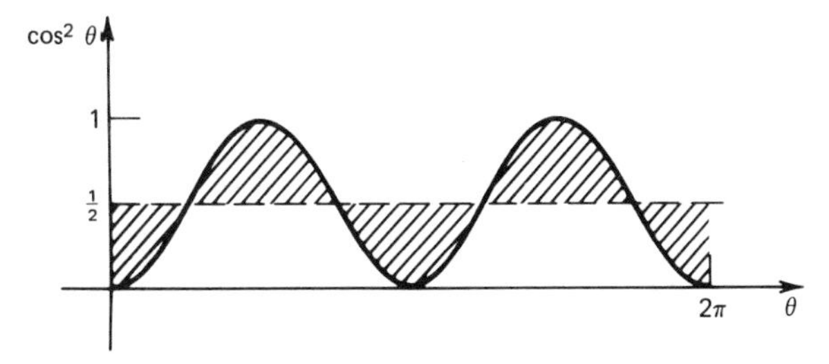

Figure 1.21. The function $\cos^2\theta$ seen as $\cos 2\theta$ plotted about the dashed line.

and since a similar argument holds for the $\sin^2\theta$, the potential energy time average over a cycle is

$$\langle \text{P.E.} \rangle = \frac{1}{2}\left(\frac{KA^2}{2}\right) = \frac{1}{2}\text{P.E.}_{\max} \tag{1.69}$$

making the total energy

$$E = \langle \text{K.E.} \rangle + \langle \text{P.E.} \rangle = \frac{m\omega^2 A^2}{2} \tag{1.70}$$

Of course, since the total energy is not time dependent a short cut to this result is secured by summing Eqs. (1.66a) and (1.66b), making use of $\sin^2\theta + \cos^2\theta = 1$, to obtain

$$E = \frac{m\omega^2 A}{2}\cos^2(\omega t + \alpha) + \frac{KA^2}{2}\sin^2(\omega t + \alpha) = \frac{m\omega^2 A^2}{2} \tag{1.71}$$

1.8 STATE OF THE SHM OPERATION

In preparation for the kind of mathematical entities brought into play in quantum theory, it is of interest here to define *operators* and *states* of the classical system. The most general solution for SHM is

$$\psi(t) = C\sin(\omega t + \alpha) \tag{1.72}$$

where $\psi(t)$ is any characteristic property obeying the equation of state

$$\left(\frac{d^2}{dt^2} + \omega^2\right)\psi(t) = 0 \tag{1.73}$$

From the preceding analyses $\psi(t)$ could be thought of as representing displacement, velocity, acceleration, and even kinetic or potential energy. However, in other systems there are numerous examples of oscillating properties such as the angular displacement of a pendulum or the charge, current, voltage, and electric and magnetic fields of sinusoidally varying electromagnetic effects. $\psi(t)$ does not need to be restricted to conventional physical properties; it may in some instances be representative of, say, a sinusoidally time-varying function whose absolute square is a probability function. In this case

$$|\psi(t)|^2 = \psi^*(t)\psi(t) \propto \text{a probability function} \tag{1.74}$$

where the asterisk denotes complex conjugate [a necessary operation if $\psi(t)$ is a complex function]. We have already seen somewhat analogous examples in that

$$\text{P.E.} \sim [x(t)]^2 \tag{1.75}$$

and

$$E_{\text{total}} \sim a^*(t)a(t)$$

Another aspect of the general function $\psi(t)$ is that it contains all the information necessary to describe the state of the system. Knowing, for example, $x(t)$ for the oscillator allows knowledge of the velocity (or momentum), acceleration, and the kinetic and potential energy. The velocity is found by taking the derivative of the displacement. We tag the derivative with the mathematical symbol $D \equiv d/dt$, called an operator, making, for example,

$$v(t) = Dx(t) = \frac{dx(t)}{dt} \tag{1.76}$$

Hence we can *operate* on the *state* to produce another state. In the case of the characteristic equation of state for SHM (1.73), we have in operator terms

$$(D^2 + \omega^2)\psi(t) = 0 \tag{1.77}$$

where $D^2 \equiv d^2/dt^2$. By further defining an operator function $\phi(D) = D^2 + \omega^2$, we can write succinctly

$$\phi(D)\psi(t) = 0 \tag{1.78}$$

the solution of which has been the central preoccupation of this chapter.

1.9 AN EXPONENTIAL SOLUTION

The general solution to this homogeneous linear differential equation of the second order (1.77) can be obtained by purely algebraic means in the form

$$z = Re^{i\omega t} \quad \text{or} \quad R^*e^{-i\omega t} \tag{1.79}$$

where $i = \sqrt{-1}$ and R or R^* (the complex conjugate) is an arbitrary complex constant. Before we investigate the nature of this solution, let us look at some properties of exponentials with imaginary powers.

Making use of an expansion in a Taylor's series (under the assumption that it can be done for imaginary quantities), we have

$$e^{i\theta} = 1 + i\theta + \frac{(i\theta)^2}{2!} + \frac{(i\theta)^3}{3!} + \cdots \tag{1.80}$$

and with the aid of $i^2 = -1$, $i^3 = -i$, and so forth is

$$e^{i\theta} = 1 + i\theta - \frac{\theta^2}{2!} - \frac{i\theta^3}{3!} + \frac{\theta^4}{4!} + \frac{i\theta^5}{5!} - \cdots$$

Rearranged into

$$e^{i\theta} = \left(1 - \frac{\theta^2}{2!} + \frac{\theta^4}{4!} - \cdots\right) + i\left(\theta - \frac{\theta^3}{3!} + \frac{\theta^5}{5!} - \cdots\right)$$

the first series in the parentheses is the expansion of $\cos\theta$ while the second represents the $\sin\theta$. Thus, we obtain Euler's relations

$$e^{i\theta} = \cos\theta + i\sin\theta \tag{1.81}$$

and by changing θ to $-\theta$ we obtain

$$e^{-i\theta} = \cos\theta - i\sin\theta \tag{1.82}$$

or conversely

$$\cos\theta = \frac{e^{i\theta} + e^{-i\theta}}{2} \tag{1.83}$$

and

$$\sin\theta = \frac{e^{i\theta} - e^{-i\theta}}{2i} = \frac{i(e^{-i\theta} - e^{i\theta})}{2}$$

The complex function in the form

$$z = Re^{i\theta} \tag{1.84}$$

has some properties we will use to advantage not only in classical wave theory but especially in the quantum description of light. For $R = R_0$, a real number, z can be looked upon as a vector in the complex plane (Fig. 1.22), having magnitude R_0 and direction specified by θ. Components (or

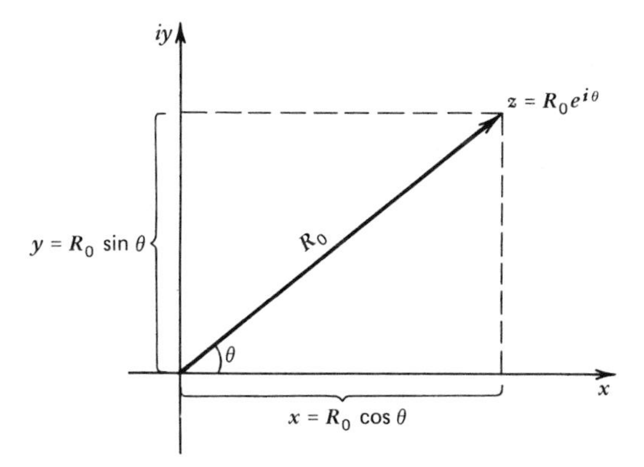

Figure 1.22. The function $e^{i\theta}$ as a vector in the complex plane.

projections) of z on the real x axis and imaginary y axis are $x = R_0 \cos\theta$ and $y = R_0 \sin\theta$, respectively, making

$$z = R_0 e^{i\theta} = x + iy \tag{1.85}$$

The complex conjugate

$$z^* = R_0 e^{-i\theta} = x - iy \tag{1.86}$$

is a function symmetric with z about the x axis making the quantity

$$zz^* = R_0^2 = x^2 + y^2 \tag{1.87}$$

If R itself is complex (as we will have occasion to use) then

$$R = x_0 + iy_0 = R_0 e^{i\alpha} \tag{1.88}$$

where x_0, y_0, R_0, and α are arbitrary real numbers related by

$$R_0^2 = x_0^2 + y_0^2 \tag{1.89}$$

and

$$\tan\alpha = \frac{y_0}{x_0}$$

which makes the function (1.84) become

$$z = R_0 e^{i(\theta+\alpha)} \tag{1.90}$$

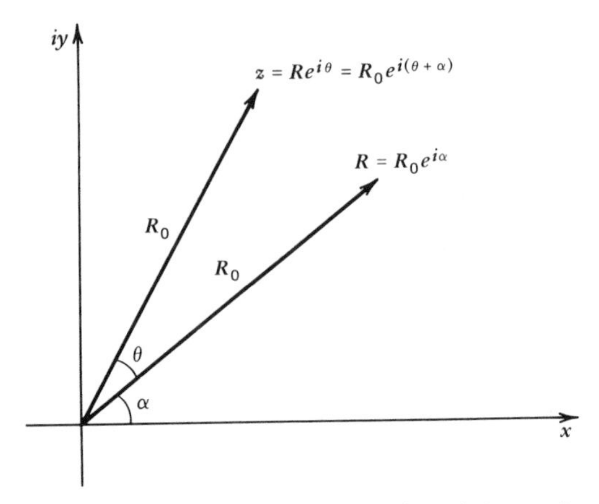

Figure 1.23. A complex amplitude R contributing a phase shift α to the complex vector.

where α has the effect of rotating the vector an additional amount in the complex plane (Fig. 1.23).

While imaginary numbers may appear in intermediate steps of a mathematical derivation used to analyze physical processes, any final observable (measurable) physical quantity cannot involve the $\sqrt{-1}$, but must be a real number. Thus if z in Eq. (1.90) is a solution satisfying the differential equation, say, for SHM, Eq. (1.79) where $\theta = \omega t$, it has physical meaning only if we take the real part of the complex function

$$\operatorname{Re} z = \operatorname{Re}(Re^{i\omega t}) = R_0 \cos(\omega t + \alpha) \tag{1.91}$$

or since the complex conjugate may also be a solution

$$\operatorname{Re} z = \tfrac{1}{2}(z + z^*) \tag{1.92}$$

The exponential form will be found particularly satisfying in describing the superposition of waves since the algebra and calculus of exponentials are simpler than those of sinusoidal functions. In quantum theory it is essential; state vectors are complex functions.

REFERENCES

Feynman, R. P., R. B. Leighton, and L. M. Sands. *The Feynman Lectures on Physics*, Vol. 1, Addison-Wesley, Reading, Mass., 1963.

Reddick, H. W. and F. H. Miller. *Advanced Mathematics for Engineers*, Wiley, New York, 1953.

Young, H. D. *Fundamentals of Waves, Optics, and Modern Physics*, McGraw-Hill, New York, 1976.

PROBLEMS

1.1. Plot the functions $\sin\theta$, $\sin^2\theta$, and $\sin(\theta^2)$. Are they all periodic functions?

1.2. Write the function $y = A\sin^2(\theta + \alpha)$ as $y' = A'\cos\theta' + B$. Find the new amplitude and phase.

1.3. Using the Taylor series for the $\sin\theta$ and $\cos\theta$, write the derivative for each function.

1.4. Superimpose the two solutions $x_1 = A\sin(\omega t + \alpha_1)$ and $x_2 = B\cos(\omega t + \alpha_2)$ for SHM. Find the amplitude and phase for this superposition.

1.5. Make a scale from $k = 0$ to $k = 10\ \text{cm}^{-1}$ on a linear axis. Label each unit $k = 1\ \text{cm}^{-1}$, $2\ \text{cm}^{-1}$, etc., also with the corresponding appropriate wavelengths λ. This scale is referred to as k-space or inverse space. Explain why.

1.6. What is the quantity ω/k? Can you give a physical explanation?

1.7. If the quantity $A\sin\omega t + B\cos\omega t = 0$ for all t, what must be true for A and B?

1.8. For an unusual spring, if the restoring force is given by $F = -Kx - K'x^2$, set up the differential equation and perform the first integration using $\ddot{x} = \frac{1}{2}(d\dot{x}^2/dx)$ and the initial conditions $x = A$ and $\dot{x} = 0$ at $t = 0$.

1.9. What is the work done by the force in Problem 1.8?

1.10. The equation of motion for a simple pendulum is $\ddot{\theta} = -\omega^2\sin\theta$. Using the first two terms of the Taylor series and the boundary conditions $\theta = \theta_0$ and $\dot{\theta} = 0$ at $t = 0$, express the solutions for $\theta(t)$ as an integral. Can this integral be found in closed form?

1.11. Perform the addition of the two complex functions $z_1 = R_1 e^{i\theta_1}$ and $z_2 = R_2 e^{i\theta_2}$ where $R_1 = x_1 + iy_1$ and $R_2 = x_2 + iy_2$. Graph the result.

1.12. If α is the complex number $\alpha = \alpha_0 + i\alpha_1$, in the expression $z = R_0 e^{i\alpha}$ where R_0, α_0, and α_1 are real numbers, find the real and imaginary parts of z. Evaluate the quantity zz^* in terms of R_0, α_0, and α_1.

2

A Wave Classic

Ultimately we wish to quantize electromagnetic waves in order to display photon properties. The formulation of that procedure will be based on the quantum harmonic oscillator. Since we have just reviewed the classical harmonic oscillator, it is useful to build upon the oscillator analysis in order to develop the notion of a classical wave.

2.1 THE CASE OF A NONWAVE

Suppose we take the ideal simple harmonic oscillator vibrating along a vertical y direction and superimpose a constant horizontal velocity v on the oscillator (Fig. 2.1*a*). This can be done in two ways (which are really one and the same): moving the entire oscillator (mass, spring, and ceiling to which it is attached) in the positive x direction with a constant velocity, or observing the fixed oscillator from an inertial coordinate system moving with a constant velocity v in the negative x direction. In either case the mass on the end of the spring will cut a trajectory, as shown in Fig. 2.1*b*. Since the vertical displacement of the oscillator is described by $y = A\cos(\omega t + \alpha)$ and the progress along x is described by $x = vt$, the equation of the trajectory is given by

$$y = A\cos\left(\omega\frac{x}{v} + \alpha\right) \tag{2.1}$$

Thus, the shape of the path is also sinusoidal, and the distance x traversed in a complete cycle of the oscillator is

$$\theta_{\text{cycle}} = \frac{\omega x}{v} = 2\pi$$

which yields the length of a cycle

$$\lambda = x_{\text{cycle}} = v\frac{2\pi}{\omega} = vT \tag{2.2}$$

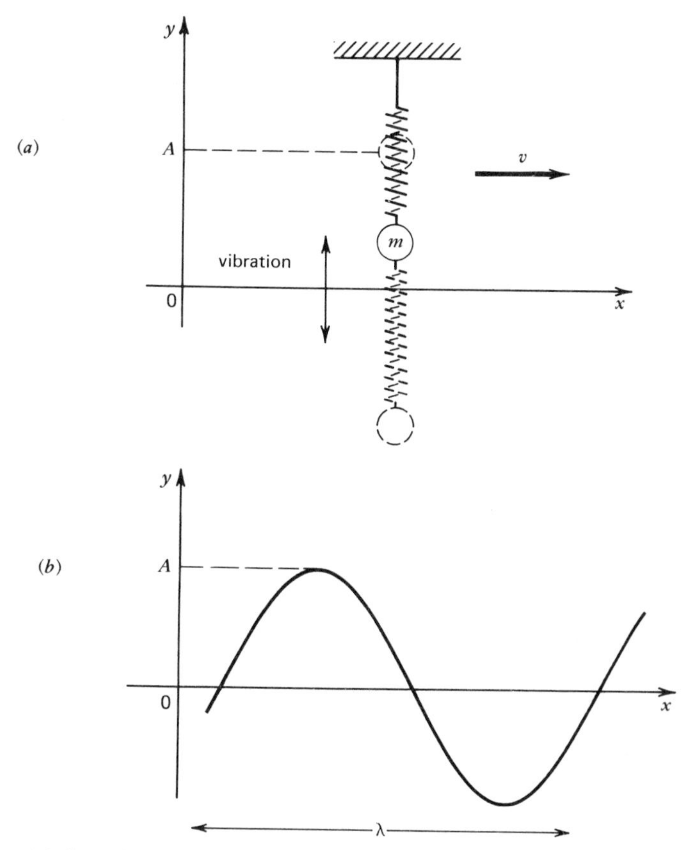

Figure 2.1. (a) A vertical oscillator with a constant horizontal velocity; (b) trajectory of the oscillator.

In other words, the distance traveled in the x direction in the time T (the time necessary to complete one cycle of oscillation which is the period) is the wavelength of this sinusoidal path. However, this is *not a traveling wave*, but merely a trajectory with a wave form or wave shape just as, for instance, a driven golf ball follows a parabolic trajectory. A traveling wave does not transport mass, but does transport energy, momentum, and angular momentum. This example does serve to establish that the velocity is related to the periodic constants λ and T, and since we have already shown $T = 2\pi/\omega$ and $\lambda = 2\pi/k$, the velocity in Eq. (2.2) is

$$v = \frac{\lambda}{T} = \frac{\omega}{k} \tag{2.3}$$

This relation as we shall see shortly will hold for traveling waves and will have a special significance.

2.2 AN IDEAL WAVE

Now let us consider a large array of identical but independent, noninteracting oscillators, each having a mass Δm separated by equal distances Δx capable of oscillating in the y direction (Fig. 2.2). If we put each oscillator separately into motion, the function

$$y_n = A_n \cos(\omega t + \alpha_n) \tag{2.4}$$

describes the displacement of the array. If we then make each amplitude the same, there is still a choice as to when or where to start each oscillator, that is, the phase. For example, if all the oscillators are started with the same phase we end up with an oscillating line of masses. On the other hand, if we set each succeeding oscillator into motion at a time Δt later than the previous one then

$$y_0 = A \cos \omega t$$

$$y_1 = A \cos \omega(t - \Delta t)$$

$$\vdots \qquad \vdots$$

$$y_n = A \cos \omega(t - n\Delta t)$$

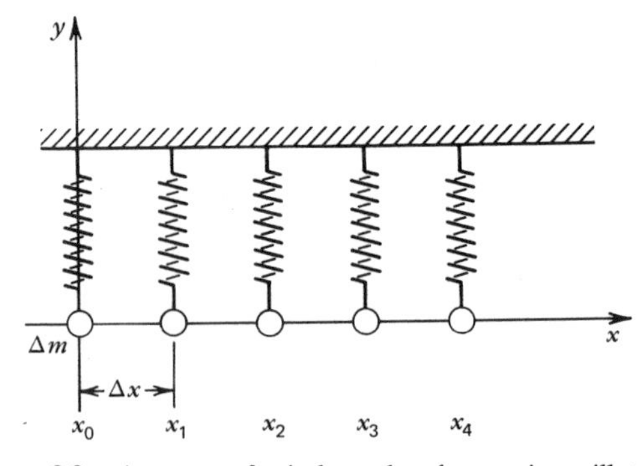

Figure 2.2. An array of n independent harmonic oscillators.

describes the displacement of the oscillators where for convenience we make $\alpha_n = 0$ for all n. We arrange the oscillators as in Fig. 2.2 so that $x_n = n\Delta x$; the time delay in starting each oscillator, $t'_n = n\Delta t$, can be written

$$t'_n = n\Delta t = \frac{x_n}{v} \tag{2.5}$$

where $v = \Delta x/\Delta t$. The equation for the nth oscillator becomes

$$y_n = A \cos \omega\left(t - \frac{x_n}{v}\right) \tag{2.6}$$

where v represents the velocity with which we move to the next oscillator to set it in motion with the same initial phase as the previous one. For this reason, v can be looked upon as a velocity for constant phase, that is, the phase velocity.

Now we wish to shrink each mass to an infinitesimal dm and distribute an infinite array of oscillators, each dx apart, located at all x, such that the above description becomes

$$y = A \cos \omega\left(t - \frac{x}{v}\right) \tag{2.7}$$

where the phase velocity $v = dx/dt = \omega/k$. Equation (2.7) is a continuous function $y(x, t)$ representing the displacement of the oscillators for all x and t; it is mathematically a traveling wave function. Moving along the x axis with a speed v, one remains at a point of constant phase and sees a static sinusoidal envelope (Fig. 2.3) in the moving (primed) coordinate system described by

$$y = A \cos(\phi - kx') \tag{2.8}$$

where ϕ is a constant. Of course, what is really traveling is the energy

$$dE = \tfrac{1}{2}(dm)\omega^2 A^2 \tag{2.9}$$

which is being injected into each successive oscillator at a time $dt = dx/v$ later than the previous one. Letting $\mu = dm/dx$ we can define a linear energy density

$$\frac{dE}{dx} = \frac{1}{2}\mu\omega^2 A^2 \tag{2.10}$$

as the energy/unit length being supplied to the system down the line. We

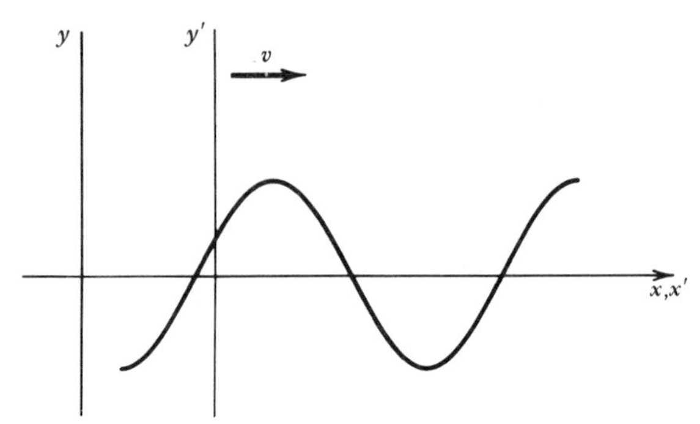

Figure 2.3. A traveling wave $y = A\cos\omega(t - x/v)$ seen from y' as a static wave $y' = A'\cos(\phi - kx')$.

must remember that these oscillators are noninteracting. This is a rather artificial way to produce a wave, but not entirely useless conceptually, when we consider the production of coherent light in a laser later.

A more convenient way to make traveling waves is to stay in one spot and wiggle only the first dm into oscillatory motion, but devise some interaction with the next dm so that it too must respond with a similar motion after a phase delay of $dt = dx/v$. In this manner we do not need to run along the wave but pour energy into the system in one place, and let it travel with the wave through the interaction. In the simplest case the interaction would have to be just right to set the next oscillator into identical motion with no distortion or energy loss. We shall discover that this ideal wave motion is difficult to accomplish.

2.3 A WAVE ON A STRING

Having fantasized the perfect wave, let us look at the somewhat more realistic situation of a continuous mass distribution as represented by a long string under tension. We will consider the mass to be a linear, uniform distribution, but neglect the fact that it is composed of atoms. The mass/unit length is μ and the tension F. At some instant we will give the string a small deflection, as in Fig. 2.4*a*, and concentrate on an element of that deflection (Fig. 2.4*b*), assuming that F and μ remain constant. The net vertical and horizontal forces on the element are given by

$$F_{y_{\text{net}}} = F\sin\theta_2 - F\sin\theta_1 \tag{2.11a}$$

and

$$F_{x_{\text{net}}} = F\cos\theta_2 - F\cos\theta_1 \tag{2.11b}$$

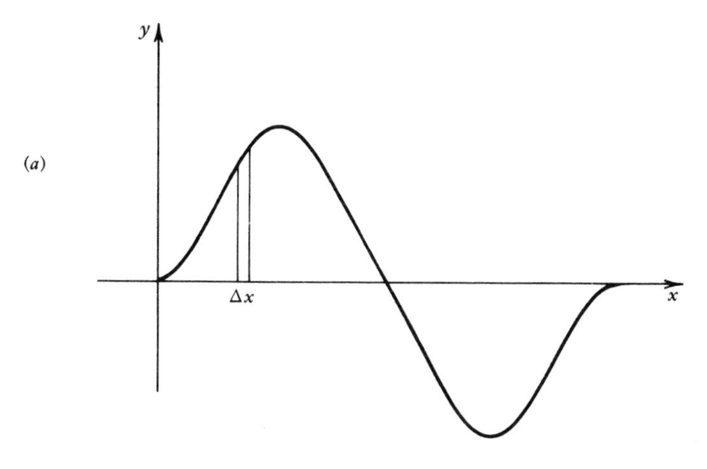

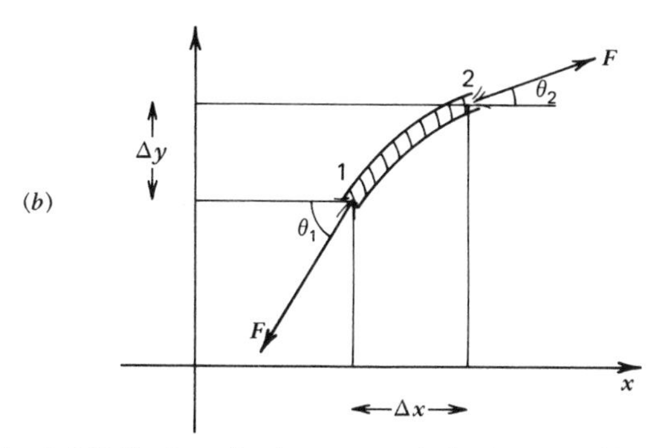

Figure 2.4. (a) Deflection of string at a particular instant; (b) an element of the deflection.

If the y displacement is truly small we can make the approximation that θ_1 and θ_2 are small, so that the sine of each angle is approximately the same as the tangent of the angle and the cosine of either angle is approximately equal to one.

Using the approximations, Eqs. (2.11a) and (2.11b) become

$$F_{y_{\text{net}}} \simeq F(\tan\theta_2 - \tan\theta_1)$$

$$F_{x_{\text{net}}} \simeq 0$$

The net vertical force is a restoring force acting opposite to the displacement of the element; that the net horizontal force is approximately zero assures no motion of the element in the x direction. Furthermore, the tangent of the angle is the slope of the curve

$$s = \frac{\partial y}{\partial x} = \tan\theta \tag{2.12}$$

so that the net vertical force becomes

$$F_{y_{\text{net}}} = F\left[\left(\frac{\partial y}{\partial x}\right)_2 - \left(\frac{\partial y}{\partial x}\right)_1\right] \tag{2.13}$$

Using Newton's second law on the mass Δm contained in the element Δx

$$F_{y_{\text{net}}} = F\left[\left(\frac{\partial y}{\partial x}\right)_2 - \left(\frac{\partial y}{\partial x}\right)_1\right] = \mu\,\Delta x\frac{\partial^2 y}{\partial t^2} \tag{2.14}$$

where we assume, in spite of the stretch, that the linear mass density $\mu = \Delta m/\Delta x$ remains constant and the entire element has a vertical acceleration $\partial^2 y/\partial t^2$.

Some of these approximations are less disturbing in approaching the limit of an infinitesimal element:

$$\lim_{\Delta x\to 0}\frac{F[(\partial y/\partial x)_2 - (\partial y/\partial x)_1]}{\Delta x} = \mu\frac{\partial^2 y}{\partial t^2} \tag{2.15}$$

In the limit, the difference in slope becomes the second derivative, and we arrive at a linear partial differential equation of the second order

$$\frac{\partial^2 y}{\partial x^2} = \frac{\mu}{F}\frac{\partial^2 y}{\partial t^2} \tag{2.16}$$

which describes the relation of the function $y(x, t)$ for any displacement in time and position for the string. With all the approximations demanded it is somewhat surprising that we arrive at any governing equation at all, the most severe demand being small amplitudes. But we do now have a mechanism for restoring forces which can lead to some kind of periodic motion.

As it stands, Eq. (2.16) professes that the coordinate rate of change of the slope of a displacement is proportional to the time rate of change of the vertical velocity. The symmetry involved is better seen if we write the equation as

$$\frac{\partial^2 y}{\partial x^2} = \frac{\partial^2 y}{\partial\left(\sqrt{F/\mu}\, t\right)^2} \tag{2.17}$$

The expression $\sqrt{F/\mu}$ is the propagation velocity v along the x axis, and since F and μ are assumed constants of the medium, the propagation velocity is constant and determined by the characteristics of the medium. Another way of viewing this wave equation is through the slope s and vertical velocity $v_y = \partial y/\partial t$, in terms of which Eq. (2.17) becomes

$$\frac{\partial s}{\partial x} = \frac{1}{v^2}\frac{\partial v_y}{\partial t} \tag{2.18}$$

and since

$$\frac{\partial^2 y}{\partial x\, \partial t} = \frac{\partial v_y}{\partial x} = \frac{\partial s}{\partial t} \tag{2.19}$$

we can generate

$$\frac{\partial^2 v_y}{\partial x^2} = \frac{1}{v^2}\frac{\partial^2 v_y}{\partial t^2} \tag{2.20a}$$

and

$$\frac{\partial^2 s}{\partial x^2} = \frac{1}{v^2}\frac{\partial^2 s}{\partial t^2} \tag{2.20b}$$

which tells us that not only the displacement y, but also the slope and vertical velocity are wave functions. The wave equation

$$\left[\frac{\partial^2}{\partial x^2} - \frac{1}{v^2}\frac{\partial^2}{\partial t^2}\right]\psi(x, t) = 0 \tag{2.21}$$

is more general than its specific derivation here for a string and will appear in many other areas of physics.

2.4 FUNCTIONS OF WAVES

Our next task is to seek solutions to the wave equation. In the form

$$\frac{\partial^2 y}{\partial x^2} = \frac{1}{v^2}\frac{\partial^2 y}{\partial t^2} = h(x, vt) \tag{2.22}$$

the function $y(x, vt)$ is such that the second partial derivative with respect to either x or vt must yield the same result, $h(x, vt)$. As with SHM we could attempt to divine the solution from known clues. For instance, the solution should be periodic in both x and vt, such that there exists some Δx and Δt allowing either

$$y(x - \Delta x, vt) = y[x, v(t + \Delta t)] \tag{2.23}$$

or

$$y[x + \Delta x, v(t + \Delta t)] = y(x, vt)$$

This condition can be satisfied by assuming solutions in the form

$$y(x, vt) = y(x \mp vt) \tag{2.24}$$

which also yield the auxiliary relations

$$\frac{\partial y}{\partial x} = \mp \frac{1}{v} \frac{\partial y}{\partial t} \tag{2.25}$$

If we consider $y(x - vt)$ as some sort of function, as in Fig. 2.5, an increment in time Δt yields $y[x - v(t + \Delta t)]$. But this is the same as $y[(x - v\,\Delta t) - vt]$. By setting $x' = x - v\,\Delta t$, the function $y(x' - vt)$ is the same function as $y(x - vt)$ as seen from a primed coordinate system moving to the right with a velocity v. The function $y(x - vt)$ is a wave traveling in the positive x direction maintaining its form or profile. To put it in terminology encountered in electromagnetic theory, we are seeing the delayed effect of a retarded signal.

Another way of looking at the wave function is by defining a variable related to phase

$$z = x - vt \tag{2.26}$$

An increment in time Δt changes the phase at x. But by traveling forward an increment $\Delta x = v\,\Delta t$ we arrive again at the same phase

$$z = x + \Delta x - v(t + \Delta t) = x - vt \tag{2.27}$$

To remain at a point of constant phase, therefore,

$$dz = 0 = dx - v\,dt \tag{2.28}$$

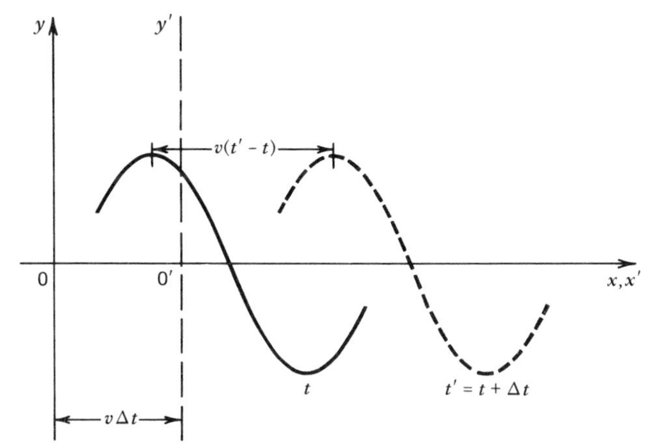

Figure 2.5. The advance of a wave in time as equivalent to the advance of the coordinate axis.

or

$$v = \frac{dx}{dt}$$

Thus v is the phase velocity previously discussed. Repeating the above arguments for $y(x + vt)$ we would find a similar wave but it would be traveling in an opposite or negative x direction.

We are referring to the solutions as waves although we have not yet shown $y(x \mp vt)$ to be any specific oscillating function. What we have shown is that the solutions to Eq. (2.22) represent a collection of y displacements along x at any particular time, such that the envelope of displacements travel along x with constant velocity. The form itself can have any shape and as we will discover, sinusoidal functions will be extremely useful for building most other wave forms.

2.5 SINUSOIDAL SOLUTIONS AND SUPERPOSITION

If we have a long string under tension on the x axis it should be possible to force the first mass element dm to move with SHM in the y direction (Fig. 2.6). Under the ideal conditions imposed in the derivation of the wave equation the next mass element should begin to move (at least theoretically) with SHM but with a phase slightly behind that of the first element. Let

$$y_1 = A \sin \omega t \tag{2.29}$$

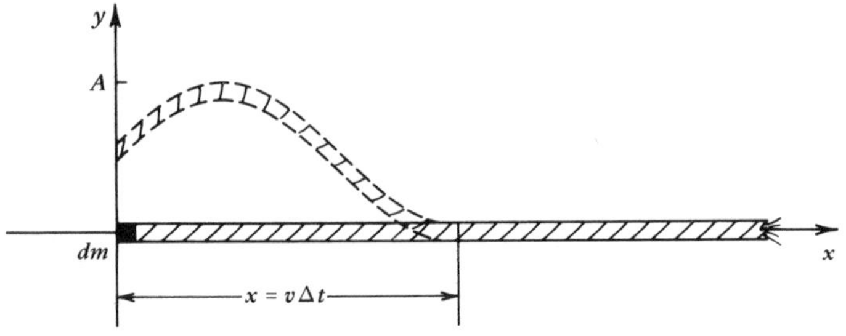

Figure 2.6. Propagation of a disturbance along a string.

describe the motion of the first element and

$$y_2 = A \sin \omega(t - dt) \tag{2.30}$$

describe the motion of the next element.

It is as if the clock for the second element starts a time dt behind the first. Since the second element is located a distance dx beyond the first, the time delay is $dt = dx/v$, where v is the speed of propagation of the disturbance. Therefore a mass element in any position x will feel the disturbance in a time $t = x/v$ and have its displacement described by

$$y = A \sin \omega\left(t - \frac{x}{v}\right) \tag{2.31}$$

where $t \geqslant x/v$. Those elements at $x > vt$ have not yet received the disturbance and hence have no displacement.

That Eq. (2.31) is a particular solution to the wave equation can be seen by taking the second partial derivatives

$$\frac{\partial^2 y}{\partial x^2} = -\frac{A\omega^2}{v^2} \sin \omega\left(t - \frac{x}{v}\right) = \frac{1}{v^2}\frac{\partial^2 y}{\partial t^2} = \frac{1}{v^2}\left[-\omega^2 A \sin \omega\left(t - \frac{x}{v}\right)\right] \tag{2.32}$$

While this is a somewhat improvised method for finding a solution to the differential equation, it will have to suffice, since there is no direct procedure for obtaining any, let alone all possible solutions. Another approach for seeking solutions to the wave equation is through the assumption of separation of variables. In this method it is assumed that

$$y(x, t) = f(x)g(t) \tag{2.33}$$

which then leads to sinusoidal functions for $f(x)$ and $g(t)$. However this assumption is as much a mathematical guess as our original intuitive demand that the solutions be in the form $y(x \mp vt)$ or its equivalent $y(t \mp x/v)$. Since the latter is more graphic because it shows clearly the role of the phase and phase velocity, it is one we here invoke to gain solutions to the wave equation.

Thus, in a more general sense, any system (not just a long string) whose physical characteristics are governed by the wave equation will admit the solutions

$$y\left(t \mp \frac{x}{v}\right) = A_{\mp} \sin(\omega t \mp kx + \alpha_{\mp}) \tag{2.34}$$

where $\omega/k = v$ is the phase velocity and α is the initial phase, with the negative and positive signs referring to waves moving in the positive or negative directions, respectively. Any linear combination of the two waves

$$f(x, t) = ay\left(t - \frac{x}{v}\right) + by\left(t + \frac{x}{v}\right) \tag{2.35}$$

where a and b are constant coefficients, will also be a solution to the differential equation. Expanding the solution for even just one of the waves, say, $y(t - x/v)$, we have

$$y\left(t - \frac{x}{v}\right) = A \sin(\omega t - kx) + B \cos(\omega t - kx) \tag{2.36}$$

where

$$A = A_{-} \cos \alpha_{-} \qquad B = A_{-} \sin \alpha_{-}$$

showing the solution to the differential equation to be a linear superposition of sinusoidal functions. If we extend this principle, not only will any linear combination of sinusoidal functions be a solution, but any linear combination of functions in the form $f(t - x/v)$ and $g(t + x/v)$ will also satisfy the equation.

In any applied situation the production of a pure single frequency sinusoidal wave is nearly impossible. Therefore, if wave theory is to be useful we must be able to design mathematical wave forms that conform closely to those that actually exist in a practical case. The superposition of waves is a method for doing just that. Through Fourier series and their extension to Fourier integrals it becomes possible to create mathematical models of a great variety of wave forms and wave packets. From this a new breed of properties will appear such as group velocity, packet spread, and transforms in k-space (wave number)—topics we deal with later.

The superposition principle also allows attempts at analyzing interference phenomena and diffraction effects. Even if the principle was useless in optics and classical wave motion (which obviously is not the case) its exploration prepares us with mathematical tools and conceptual insights that are essential to the study of quantum theory. Yet the innocence of the principle is outstanding in that it merely asserts that wave functions occupying the same place at the same time simply add. In the next chapter we investigate this addition of waves.

Before taking leave of our discussion of superposition, we will use the principle to demonstrate another breed of solutions. We have seen from Eqs. (2.35) and (2.36) that any linear combination of $\sin(\omega t \mp kx)$ and $\cos(\omega t \mp kx)$ is a solution to the wave equation. Mathematically there are no restrictions on the arbitrary constants. Choosing the constants to be real and imaginary in the combination

$$y = B\cos(\omega t \mp kx) + iA\sin(\omega t \mp kx) \tag{2.37}$$

we generate the exponential solutions

$$y = Ce^{i(\omega t \mp kx)} \tag{2.38}$$

where C is a complex coefficient

$$C = Re^{i\alpha} \tag{2.39}$$

with

$$R = (B^2 + A^2)^{1/2}$$

and

$$\alpha = \tan^{-1}\frac{A}{B}$$

The complex conjugate

$$y^* = C^*e^{-i(\omega t \mp kx)} \tag{2.40}$$

is also a solution to the wave equation. In any application to an observable property in classical physics, it is only the real part of the exponentials, Re y, that have physical significance. Their handiness, however, will be appreciated for the addition of waves where the exponential is easier to manipulate than their sinusoidal counterparts. In quantum physics these complex functions will be not only useful but required by the nature of the theory.

2.6 ENERGY OF A WAVE

Returning to the mass element in Fig. 2.4*b* for which the wave equation was formulated, we would like to examine its energy content. At the instant the mass element is in a general displacement y it possesses both kinetic and potential energy—kinetic energy since the mass $\Delta m = \mu\,\Delta x$ has a velocity dy/dt, and potential energy in that work is done on the element in raising it from its undisplaced position at an earlier time at $y = 0$. The kinetic energy is easily written as

$$K = \frac{1}{2}\Delta m\left(\frac{\partial y}{\partial t}\right)^2 = \frac{1}{2}\mu\left(\frac{\partial y}{\partial t}\right)^2 \Delta x \tag{2.41}$$

or in terms of energy density

$$U_K = \frac{\partial K}{\partial x} = \frac{1}{2}\mu\left(\frac{\partial y}{\partial t}\right)^2 \tag{2.42}$$

The potential energy is represented by the work done in displacing the element from its flat position at $y = 0$ to its position at y. If we assume as before that the tension is F and remains constant (the displacements are small), the net force required in the y direction (Fig. 2.7) is

$$F_{y_{\text{net}}} = F\left[\left(\frac{\partial y}{\partial x}\right)_{x+\Delta x} - \left(\frac{\partial y}{\partial x}\right)_x\right] \tag{2.43}$$

making the potential energy

$$P = \int_0^y F\left[\left(\frac{\partial y}{\partial x}\right)_{x+\Delta x} - \left(\frac{\partial y}{\partial x}\right)_x\right] dy \tag{2.44}$$

Defining a potential energy density as $\partial P/\partial x$ and noting that

$$\lim_{\Delta x\to 0} \frac{(\partial y/\partial x)_{x+\Delta x} - (\partial y/\partial x)_x}{\Delta x} = \frac{\partial(\partial y/\partial x)}{\partial x} = \frac{\partial(\partial y/\partial x)}{\partial y}\frac{\partial y}{\partial x}$$

the result becomes

$$U_P = \frac{\partial P}{\partial x} = \frac{F}{2}\int_0^{\left(\frac{\partial y}{\partial x}\right)^2} \partial\left(\frac{\partial y}{\partial x}\right)^2 = \frac{F}{2}\left(\frac{\partial y}{\partial x}\right)^2 \tag{2.45}$$

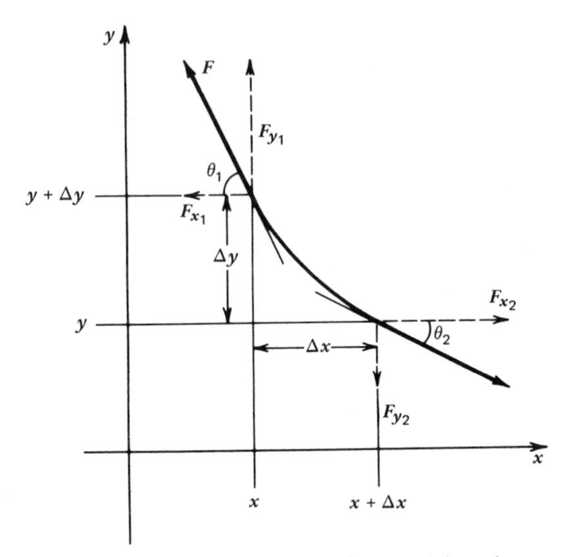

Figure 2.7. Forces acting on an element of the string.

The total energy/unit length is

$$U = U_K + \dot{U}_P = \frac{1}{2}\left[\mu\left(\frac{\partial y}{\partial t}\right)^2 + F\left(\frac{\partial y}{\partial x}\right)^2\right] \tag{2.46}$$

Since we have already shown in Eq. (2.25) that $\partial y/\partial x = (1/v)\partial y/\partial t$, the energy density can succinctly be expressed by either

$$U = \mu\left(\frac{\partial y}{\partial t}\right)^2 \quad \text{or} \quad \mu v^2\left(\frac{\partial y}{\partial x}\right)^2 \tag{2.47}$$

where we have used $\mu = F/v^2$. It is important to notice that the energy density is not constant but varies with position and time, and, of crucial importance to classical wave theory, the energy is dependent upon the square of the transverse velocity and slope.

In a particular application for a sinusoidal wave, $y = A\sin(\omega t - kx)$, the partial derivatives become

$$\frac{\partial y}{\partial t} = \omega A\cos(\omega t - kx) \tag{2.48}$$

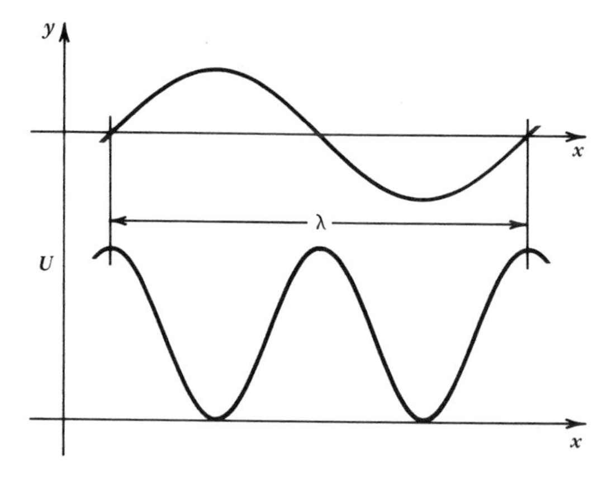

Figure 2.8. Variation in energy density for a sinusoidal wave.

and

$$\frac{\partial y}{\partial x} = -kA\cos(\omega t - kx)$$

making the energy density (Fig. 2.8)

$$U = \mu\omega^2 A^2 \cos^2(\omega t - kx) = \mu v^2 k^2 A^2 \cos^2(\omega t - kx) \tag{2.49}$$

A time average of the energy density over one cycle is

$$\langle U \rangle_{\text{time}} = \frac{\int_0^T U\,dt}{T} = \frac{\mu\omega^2 A^2}{T}\int_0^T \cos^2(\omega t - kx)\,dt \tag{2.50}$$

Since $\cos^2\theta = \frac{1}{2} + (\cos 2\theta)/2$ and the integral of a sinusoidal function over any number of complete cycles is zero, the integral yields $T/2$. The time average is half the maximum value.

$$\langle U \rangle_{\text{time}} = \tfrac{1}{2}\mu\omega^2 A^2 \tag{2.51}$$

The total energy residing in a full cycle of the wave is

$$E = \mu\int_0^\lambda \left(\frac{\partial y}{\partial t}\right)^2 dx = \mu\omega^2 A^2 \int_0^\lambda \cos^2(\omega t - kx)\,dx \tag{2.52}$$

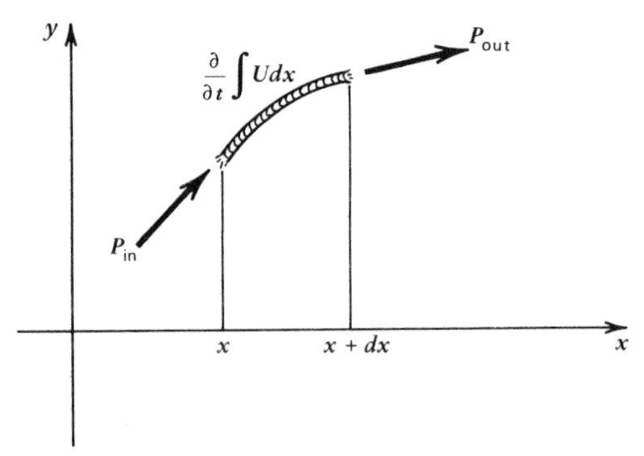

Figure 2.9. Energy flow through a segment of a string.

which upon integrating becomes

$$E = \mu\omega^2 A^2 \frac{\lambda}{2} = \pi\mu v\omega A^2 \tag{2.53}$$

By now we can see that almost any expression of energy involves a quadratic amplitude.

The energy contained in a wave between x and $x + dx$ is

$$dE = U\,dx \tag{2.54}$$

In a time $dt = dx/v$ the energy will have propagated through dx to the position $x + dx$. By defining a flux $P(x, t)$ as the rate of change of energy (power) we have

$$P(x, t) = \frac{dE}{dt} = \frac{dE}{dx}\frac{dx}{dt} = Uv \tag{2.55}$$

Since the energy density is not constant, the flux into a segment less the flux out must account for the rate of change of energy within that segment (Fig. 2.9).

$$P_{\text{in}}(x, t) - P_{\text{out}}(x + dx, t) = \frac{\partial}{\partial t}\int U\,dx \tag{2.56}$$

2.7 PLANE TRAVELING WAVES

So far we have considered waves propagating along an x axis only, where the undulating physical property was the y displacement of the system. We have also shown [Eqs. (2.20a) and (2.20b)] that the slope and y velocity in the case of a string under tension are described by the wave equation with solutions in the form $f(t \mp x/v)$. As a matter of fact if we had chosen to analyze sound waves, the fluctuating property might have been the coordinate position of a layer of air in the direction of propagation, or simply the pressure or density. In electromagnetic waves the fluctuating quantities will be electric or magnetic fields or their potentials. The common thread is that any physical variable regulated by

$$\left(\frac{\partial^2}{\partial x^2} - \frac{1}{v^2}\frac{\partial^2}{\partial t^2}\right)f = 0 \tag{2.57}$$

has the solutions $f(t \mp x/v)$ representing traveling waves in a one-dimensional coordinate system. We wish to generalize the theory to three-dimensional space.

The wave, as it has been considered until now, has been describing an infinitesimally thin string with no cross-sectional area. If the string, for example, had an extension in a z direction (Fig. 2.10) it would support waves that would appear like a sheet flapping on a clothesline and still be described by the same functions. We could call it a line traveling wave since it would be propagating perpendicular to a line, in this case, a line parallel to the z axis.

Now if the string also had an extension in the y direction (Fig. 2.11) and each point of the cross-sectional plane of the string varied with x and t in

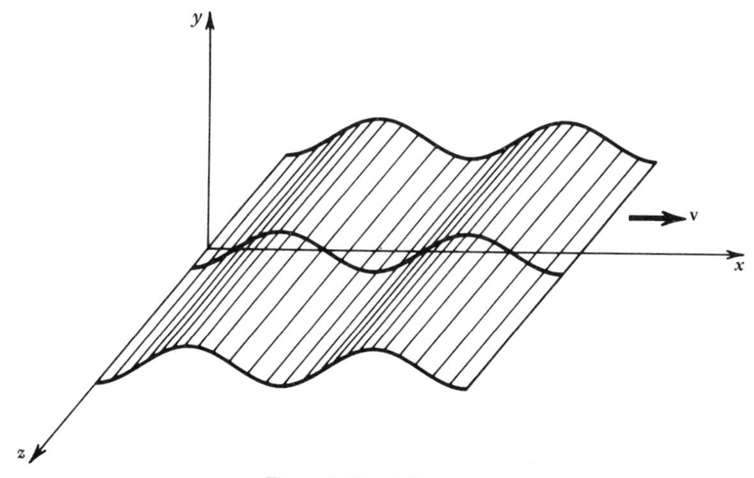

Figure 2.10. A line wave.

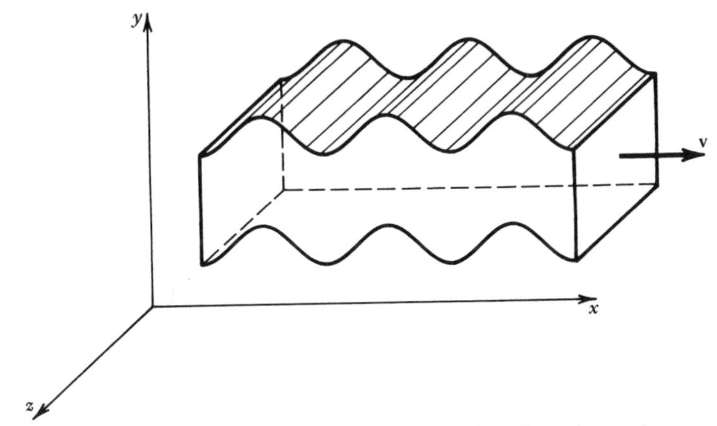

Figure 2.11. A plane traveling wave propagating along the x axis.

the same way, the wave would be a bundle moving perpendicular to the yz plane. It is in this sense that it is called a plane wave.

So far the wave is still propagating in the x direction and nothing has changed. We wish now to describe a plane wave moving in a direction along **r** in a three-dimensional coordinate system (Fig. 2.12). We define a unit vector **n** which is in the direction of propagation of the wave. The differential equation is yet one-dimensional as

$$\left[\frac{\partial^2}{\partial r^2} - \frac{1}{v^2}\frac{\partial^2}{\partial t^2}\right] f = 0 \tag{2.58}$$

with the solution

$$f = f\left(t \mp \frac{\mathbf{n}\cdot\mathbf{r}}{v}\right) \tag{2.59}$$

where $\mathbf{n}\cdot\mathbf{r} = r$. But if we wish to express the wave equation in terms of x, y, and z, we must carry out a transformation of axes.

The vector **n** is a directional unit vector specified by its direction cosines.

$$\mathbf{n} = \hat{\mathbf{i}}\cos\alpha + \hat{\mathbf{j}}\cos\beta + \hat{\mathbf{k}}\cos\gamma \tag{2.60a}$$

making

$$r = \mathbf{n}\cdot\mathbf{r} = x\cos\alpha + y\cos\beta + z\cos\gamma \tag{2.60b}$$

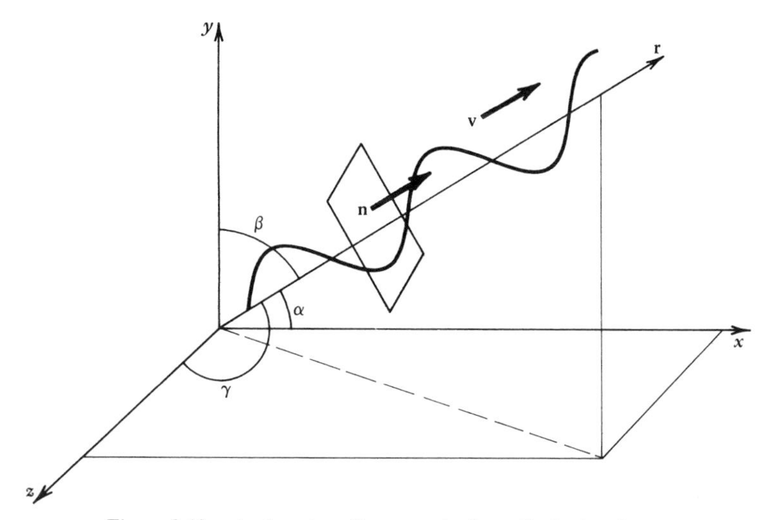

Figure 2.12. A plane traveling wave in three-dimensional space.

where

$$\cos^2\alpha + \cos^2\beta + \cos^2\gamma = 1$$

a well-known property of the direction cosines representing the Pythagorean theorem. The transformation provides

$$\frac{\partial f}{\partial x} = \frac{\partial f}{\partial r}\frac{\partial r}{\partial x} = \frac{\partial f}{\partial r}\cos\alpha \tag{2.61}$$

and

$$\frac{\partial^2 f}{\partial x^2} = \frac{\partial}{\partial x}\left(\frac{\partial f}{\partial r}\right)\cos\alpha = \frac{\partial^2 f}{\partial r^2}\cos^2\alpha$$

Doing the same for y and z we obtain

$$\frac{\partial^2 f}{\partial x^2} + \frac{\partial^2 f}{\partial y^2} + \frac{\partial^2 f}{\partial z^2} = \frac{\partial^2 f}{\partial r^2}(\cos^2\alpha + \cos^2\beta + \cos^2\gamma) = \frac{\partial^2 f}{\partial r^2} \tag{2.62}$$

which may be expressed more economically with the aid of "del" the vector operator

$$\nabla = \hat{\mathbf{i}}\frac{\partial}{\partial x} + \hat{\mathbf{j}}\frac{\partial}{\partial y} + \hat{\mathbf{k}}\frac{\partial}{\partial z} \tag{2.63a}$$

With the dot product

$$\nabla \cdot \nabla = \nabla^2 = \frac{\partial^2}{\partial x^2} + \frac{\partial^2}{\partial y^2} + \frac{\partial^2}{\partial z^2} \tag{2.63b}$$

the wave equation in terms of this Laplacian operator becomes

$$\nabla^2 f = \frac{1}{v^2}\frac{\partial^2 f}{\partial t^2} \tag{2.64}$$

with the appropriate solutions

$$f(x, y, z, t) = f\left(t \mp \frac{\mathbf{n}\cdot\mathbf{r}}{v}\right) \tag{2.65}$$

In the case of sinusoidal solutions

$$f = C\sin(\omega t \mp k\,\mathbf{n}\cdot\mathbf{r}) \tag{2.66}$$

the term containing the wave number k suggests a vector description. It becomes convenient to define a propagation vector

$$\mathbf{k} = \mathbf{n}k = k_x\hat{\mathbf{i}} + k_y\hat{\mathbf{j}} + k_z\hat{\mathbf{k}} \tag{2.67}$$

so that the wave number is

$$k = |\mathbf{k}| = \left(k_x^2 + k_y^2 + k_z^2\right)^{1/2} \tag{2.68}$$

in terms of which the plane wave becomes

$$f = C\sin(\omega t \mp \mathbf{k}\cdot\mathbf{r}) \tag{2.69}$$

where $\mathbf{k}\cdot\mathbf{r}$ can also be written as $2\pi r/\lambda$. The plane of constant phase, usually referred to as the wave front, has its normal in the $\mathbf{k}$ direction and propagates with a velocity $v_{\text{phase}} = \omega/k$.

The energy formulation in three dimensions undergoes a slight change. With a plane cross-section now available to the traveling wave, what before were linear energy densities now become energy/unit volume V. A wave propagating in the $\mathbf{r}$ direction through a cross-sectional area A has an energy density

$$U = \frac{dE}{dV} \tag{2.70}$$

In three-dimensional space the flux

$$S = vU \tag{2.71}$$

is the energy per unit time per unit area, and can henceforth be named intensity.

2.8 SPHERICAL WAVES

Another handy plane of constant phase is the spherical surface. Waves emanating from a theoretical point source uniformly in all directions would have this spherical wave front. We can gain insight to the solution by appealing to experimental evidence regarding spherical waves. The intensity of radiant energy from more or less point sources is found to drop off as the reciprocal of the distance squared from the source.

$$S \sim \frac{1}{r^2} \tag{2.72}$$

Accepting this as gospel, and knowing that classical wave intensity depends on the square of the amplitude, it would appear logical to assume that a spherical plane wave would have a form

$$\psi(r, t) = \frac{1}{r} f\left(t \mp \frac{r}{v}\right) \tag{2.73a}$$

The logic is really impeccable since, while the intensity falls away by $1/r^2$, the spherical surface area around the point source increases by r^2 keeping the total energy flow constant but spread over a larger surface (Fig. 2.13). For the case of a sinusoidal function

$$\psi(r, t) = \frac{A}{r}\cos(\omega t \mp kr) \tag{2.74}$$

the intensity is

$$S \sim [\psi(r, t)]^2 \sim \frac{A^2}{r^2}\cos^2(\omega t \mp kr) \tag{2.75}$$

Note that k and r are not vectors since this is a wave propagating uniformly in all radial directions. As before, the negative sign is for the outward expanding wave front, and the positive sign the inward moving wave. The

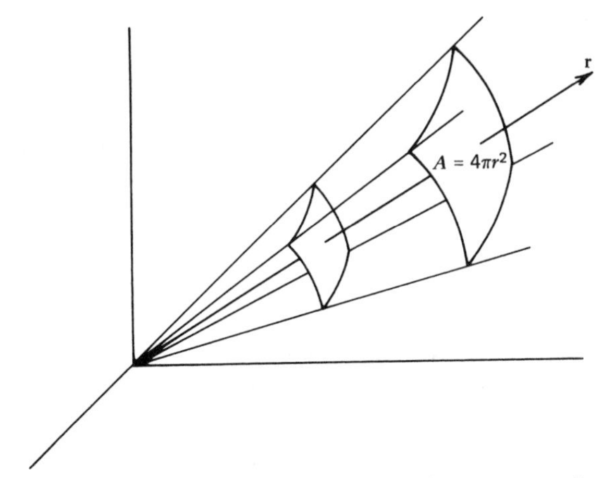

Figure 2.13. Dependence of spherical surface area on radius.

latter is difficult to obtain physically as it would necessitate a uniform spherical source producing waves imploding to a central point.

If we ponder the expression

$$r\psi(r,t) = f\left(t \mp \frac{r}{v}\right) \tag{2.73b}$$

inverse reasoning leads us to assume that $r\,\psi(r,t)$ should satisfy the differential equation

$$\left[\frac{\partial^2}{\partial r^2} - \frac{1}{v^2}\frac{\partial^2}{\partial t^2}\right] r\,\psi = 0 \tag{2.76}$$

and indeed we do arrive at precisely this result by transforming Eq. (2.64) into radial coordinates.

From Fig. 2.14, it is evident that

$$r^2 = x^2 + y^2 + z^2 \tag{2.77}$$

Transforming the first space derivative of ψ from the wave equation

$$\nabla^2\psi = \frac{1}{v^2}\frac{\partial^2\psi}{\partial t^2} \tag{2.78}$$

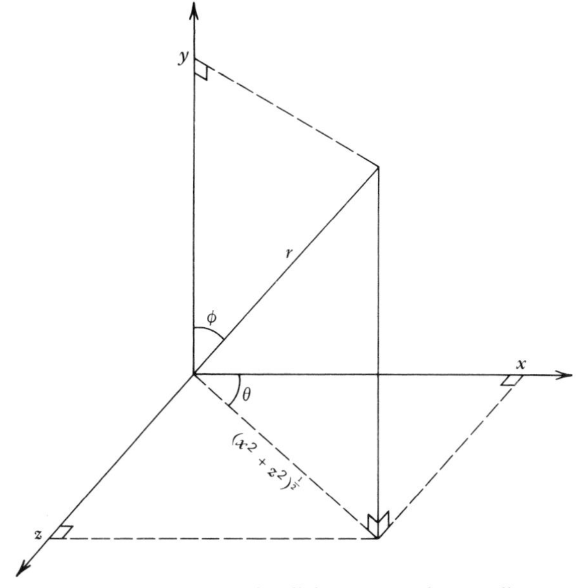

Figure 2.14. Relation of radial to rectangular coordinates.

we have

$$\frac{\partial \psi}{\partial x} = \frac{\partial \psi}{\partial r} \frac{\partial r}{\partial x} \tag{2.79}$$

With the aid of

$$2r \frac{\partial r}{\partial x} = 2x \tag{2.80}$$

from Eq. (2.77) we obtain

$$\frac{\partial \psi}{\partial x} = \frac{\partial \psi}{\partial r} \frac{x}{r} \tag{2.81}$$

The second derivative becomes

$$\begin{aligned} \frac{\partial^2 \psi}{\partial x^2} &= \frac{1}{r} \frac{\partial \psi}{\partial r} + x \left(\frac{1}{r} \frac{\partial^2 \psi}{\partial r^2} - \frac{1}{r^2} \frac{\partial \psi}{\partial r} \right) \frac{\partial r}{\partial x} \\ &= \frac{1}{r} \frac{\partial \psi}{\partial r} + \frac{x^2}{r^2} \frac{\partial^2 \psi}{\partial r^2} - \frac{x^2}{r^3} \frac{\partial \psi}{\partial r} \end{aligned} \tag{2.82}$$

When we perform a similar operation for $\partial^2\psi/\partial y^2$ and $\partial^2\psi/\partial z^2$ the Laplacian becomes

$$\nabla^2\psi = \frac{3}{r}\frac{\partial\psi}{\partial r} + \left(\frac{x^2+y^2+z^2}{r^2}\right)\frac{\partial^2\psi}{\partial r^2} - \left(\frac{x^2+y^2+z^2}{r^3}\right)\frac{\partial\psi}{\partial r} \tag{2.83}$$

which simplifies to

$$\nabla^2\psi = \frac{2}{r}\frac{\partial\psi}{\partial r} + \frac{\partial^2\psi}{\partial r^2} = \frac{1}{r}\frac{\partial^2(r\psi)}{\partial r^2} \tag{2.84}$$

Thus we arrive at what we have already presumed

$$\frac{\partial^2(r\psi)}{\partial r^2} = \frac{1}{v^2}\frac{\partial^2(r\psi)}{\partial t^2} \tag{2.85}$$

where v is the propagation velocity of spherical surfaces of constant phase.

For any point source either very far away, such as a star, or just far compared with the length of the apparatus for an observation, the $1/r$ dependence of the spherical wave function may be considered approximately constant, making a spherical wave appear to be approximately a parallel plane wave.

2.9 A WORD ON WAVES

While we have used ψ as a scalar function to represent the physical property being displaced or fluctuating in space and time, the wave equation as encountered in electromagnetic theory, will work just as well for vector functions. The introduction of this vector nature will also allow us to deal with polarization properties of waves.

As there are no precise point sources, neither are there perfectly parallel waves. Thus the application of these idealized waves must proceed with caution. We have already intimated that a pure monochromatic wave is difficult to come by in nature. Control over not only monochromaticity but also phase and duration of a wave will be subject to experimental frailty and ultimately more severely to quantum requirements.

Mechanical waves present problems enough; electromagnetic waves will prove even more harrowing. Our more usual Gaussian light sources are very sloppy indeed. They are extended in space, radiate in random phases, and always present a spread in frequencies. Doppler broadening, thermal noise, and diffraction will plague the analysis of emission, propagation, and absorption characteristics of this incoherent radiation. This is all compounded by the extremely high frequencies of the optical range [say, approximately 10^{13}–10^{18} Hz, although only $(4–7.5)\times 10^{14}$ Hz is visible light] which discourages all hope of directly measuring the amplitude variation of electromagnetic waves. In light of these difficulties, it is surprising that so much has been discovered about the nature of electromagnetic waves.

Still, it is a complex area, and much is yet to be discerned. The refinement of lasers and photodetectors has added facility to the approach of more detailed and accurate experiments, and has especially enhanced interest in the quantum optics of photons—effects in which photons can be created in coherent emission and seen almost individually in sensitive detection devices. In the region of low-intensity effects the quantum characteristics of radiant energy become important and lend interpretation to the particle view of light, a view from which the wavelike nature is a natural consequence. We are approaching an introduction to this theory. Since it is based upon knowledge of the harmonic oscillator, waves, and electromagnetic theory we will deal still further with these classical aspects in order to gain a firm footing with which we will plug into the quantum realm.

Of great use to quantum theory is the superposition principle. With it we can build more complex wave forms and gain insight into the interference effects that are germane to not only classical but quantum optics as well. Hence, we investigate the addition of waves.

REFERENCES

Ditchburn, R. W. *Light*, Interscience, New York, 1955.

Goble, A. T. and D. K. Baker. *Elements of Modern Physics*, Ronald, New York, 1962.

Meyer-Arendt, J. R. *Introduction to Classical and Modern Optics*, Prentice-Hall, Englewood Cliffs, N.J., 1972.

Rayleigh, J. W. S. *The Theory of Sound*, Vol 1., Dover, New York, 1945.

Young, H. D. *Fundamentals of Waves, Optics and Modern Physics*, McGraw-Hill, New York, 1976.

PROBLEMS

2.1. Show that $y_1 = A\cos\omega(t - x/v)$ is the same wave as $y_2 = A\cos\omega(x/v - t)$. Then show that $y_3 = A\cos\omega(t + x/v)$ is the same as y_1 but moving in the opposite direction.

2.2. Find the combination of traveling waves that represents each of the four wave functions: $\sin\omega t\sin kx$, $\sin\omega t\cos kx$, $\cos\omega t\sin kx$, $\cos\omega t\cos kx$.

2.3. Show that $f(x, y, t) = A \cos \omega[t - (x \cos \theta + y \sin \theta)/v]$ is a line wave traveling at an angle θ to the x axis.

2.4. Assuming a solution to the wave equation in the form $y(x, t) = f(x)g(t)$, find $f(x)$ and $g(t)$ for the general solution. How is $y(x, t)$ equivalent to a traveling wave?

2.5. Demonstrate that $y = Ae^{\pm i(\omega t \mp kx)}$ are solutions of the wave equation.

2.6. Show that $f(x, t) = Ae^{-\sigma^2(x - vt)^2}$ is a solution to the wave equation. Plot the wave at a particular time t and discuss its shape.

2.7. Is the function $f(x, t) = A \cos 2\pi(x/\lambda + t/T + \alpha)$ a solution to the wave equation? Explain the constants.

2.8. Using the exponential solutions, write the energy expression for a traveling wave.

2.9. From Fig. 2.12 show the components of the vector $\mathbf{r}$, and prove that $\cos^2\alpha + \cos^2\beta + \cos^2\gamma = 1$.

2.10. Perform the operations to show that

$$\frac{1}{r} \frac{\partial^2}{\partial r^2}(r \psi) = \frac{2}{r} \frac{\partial \psi}{\partial r} + \frac{\partial^2 \psi}{\partial r^2}$$

3

Sums, Series, and Packets

Unlike massive particles, waves can occupy the same place at the same time. The combined effect of two or more wave disturbances at any given space–time point $(\mathbf{r}, t)$ is the straightforward sum or superposition of the individual wave displacements. The resultant field of displacements, by this classical wave picture, provides the basis for predicting patterns of energy flow encountered in such events as the interference and detection of waves.

We look upon light as a wave phenomenon. The transport of optical energy is conducted by means of electromagnetic waves, whose nature is considered in subsequent chapters. For the present, it suffices to know that electromagnetic waves are represented by the same mathematical tools we are evolving for transverse traveling waves. It is our purpose then to explore the superposition of elemental waves in order to form wave groups and packets with which we can further describe the propagation and distribution of radiant energy.

3.1 ADDITION OF SIMILAR SINUSOIDAL WAVES

Temporarily ignoring the physical impossibility of producing perfect sinusoidal waves, we assume their mathematical existence and perform a variety of combinations with them. With the sinusoidal solution to the wave equation in the form

$$y = A\cos(\omega t - kx + \phi) \tag{3.1}$$

the sum of two such monochromatic (single frequency) waves traveling in the same direction, differing only in phase and amplitude (Fig. 3.1) is

$$y = y_1 + y_2 = A_1\cos(\theta + \phi_1) + A_2\cos(\theta + \phi_2) \tag{3.2}$$

where $\theta = \omega t - kx$. With the aid of the trigonometric expansions

$$\cos(\theta \pm \phi) = \cos\theta\cos\phi \mp \sin\theta\sin\phi \tag{3.3}$$

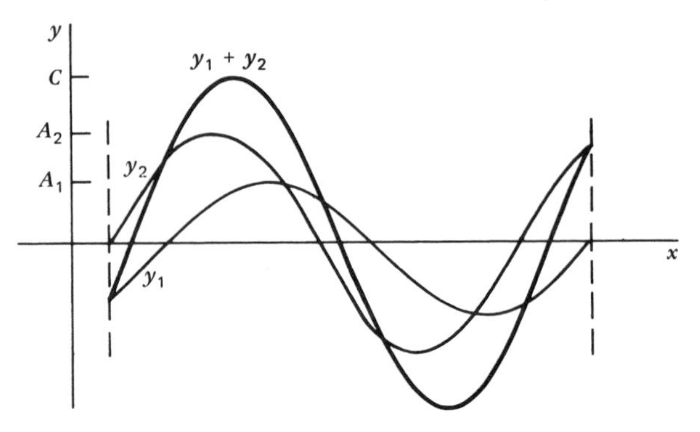

Figure 3.1. The sum of two sinusoidal waves.

the sum becomes

$$y = (A_1 \cos \phi_1 + A_2 \cos \phi_2) \cos \theta - (A_1 \sin \phi_1 + A_2 \sin \phi_2) \sin \theta \quad (3.4)$$

When two new constants are identified as

$$C \cos \delta = A_1 \cos \phi_1 + A_2 \cos \phi_2 \quad (3.5)$$

and

$$C \sin \delta = A_1 \sin \phi_1 + A_2 \sin \phi_2$$

the superposition of the two waves becomes

$$y = C \cos(\omega t - kx + \delta) \quad (3.6)$$

where

$$C^2 = (A_1 \cos \phi_1 + A_2 \cos \phi_2)^2 + (A_1 \sin \phi_1 + A_2 \sin \phi_2)^2 \quad (3.7)$$

and

$$\tan \delta = \frac{A_1 \sin \phi_1 + A_2 \sin \phi_2}{A_1 \cos \phi_1 + A_2 \cos \phi_2} \quad (3.8)$$

Thus, any two sinusoidal functions superimpose into yet another sinusoidal function, differing from the original components only in amplitude and phase. As a matter of fact, the definitions (3.5) of the resultant amplitude and phase, C and δ, suggest a vector sum of component properties of the original amplitudes and phases. Equation (3.7) for the resultant amplitude, upon expansion, is

$$C^2 = A_1^2(\sin^2\phi_1 + \cos^2\phi_1) + A_2^2(\sin^2\phi_2 + \cos^2\phi_2)$$

$$+2A_1A_2(\cos\phi_1\cos\phi_2 + \sin\phi_1\sin\phi_2) \tag{3.9}$$

which is none other than the law of cosines:

$$C^2 = A_1^2 + A_2^2 + 2A_1A_2\cos(\phi_1 - \phi_2) \tag{3.10}$$

This vector nature for the addition of waves will appear clearly when we perform the same summation as above but in an exponential formulation. In the meantime, let us note some generalizations and two unique extreme cases for this superposition.

1. If the phase difference between the two waves y_1 and y_2 is such that the waves are in phase, $\delta = 0$, the resultant amplitude has its largest possible value for the addition of two waves. The composite wave seen in Fig. 3.2*a* is

$$y = C\cos(\omega t - kx + \delta) \tag{3.11}$$

where $\delta = \phi_1 = \phi_2$ and $C = A_1 + A_2$, the maximum amplitude.

2. Should the phase of the component waves differ by π (or any odd multiple thereof), the composite function in Fig. 3.2*b* is

$$y = A_1\cos(\omega t - kx + \phi_1) + A_2\cos[\omega t - kx + (\phi_1 - \pi)] \tag{3.12}$$

or

$$y = (A_1 - A_2)\cos(\omega t - kx + \phi_1) \tag{3.13}$$

making the composite amplitude $C = A_1 - A_2$ a minimum. In the special case of equal component amplitudes, namely, $A_1 = A_2$, the resultant function vanishes (Fig. 3.2*c*).

The principle of superposition is not limited to just two sinusoidal waves, but applies to the sum of any number of components.

$$y = \sum_i y_i = \sum_i A_i\cos(\omega t - kx + \phi_i) \tag{3.14}$$

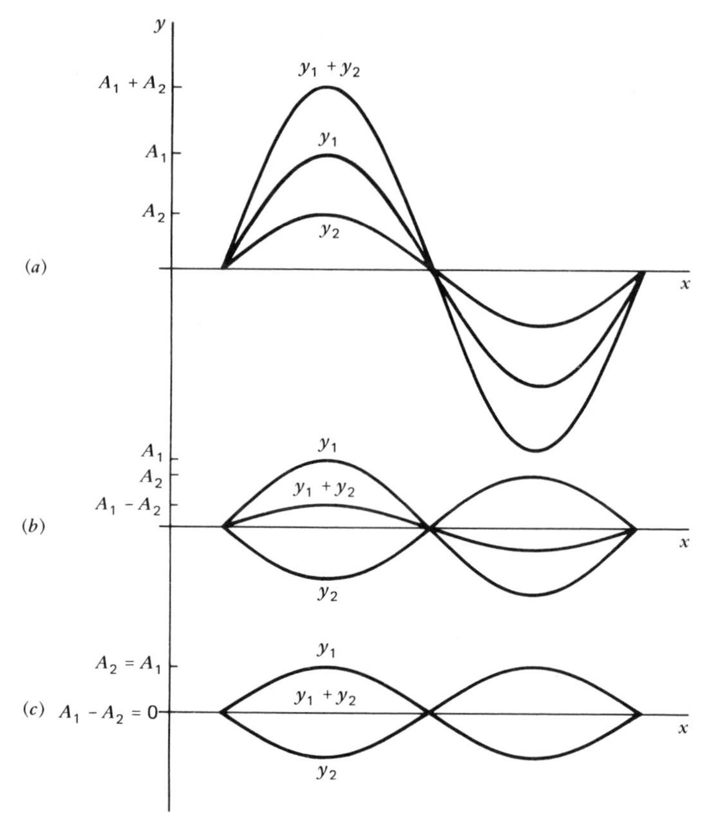

Figure 3.2. The sum of two waves for (a) equal phase; (b) opposite phase; (c) opposite phase and equal amplitude.

The generalization for handling this summation of similar traveling waves of identical frequency and wavelength follows from the definitions of Eqs. (3.5). When we define the constants

$$C\cos\delta = \sum_i A_i \cos\phi_i \quad \text{and} \quad C\sin\delta = \sum_i A_i \sin\phi_i \tag{3.15}$$

the sum over n sinusoidal components yields

$$y = C\cos(\omega t - kx + \delta) \tag{3.16}$$

where

$$C^2 = \left(\sum_{i=1}^{n} A_i \cos\phi_i\right)^2 + \left(\sum_{i=1}^{n} A_i \sin\phi_i\right)^2$$

and

$$\tan \delta = \frac{\sum_{i=1}^{n} A_i \sin \phi_i}{\sum_{i=1}^{n} A_i \cos \phi_i}$$

thus demonstrating that any number of sinusoidal waves in superposition constitute still another sinusoidal wave. Rhetorically, we can wonder whether the components even exist. The addition process is similar to the idea that any set of linear vectors can be composed into a single resultant vector, the components of which are a convenient (but not necessarily unique) basis. This sense of superposition will appear repeatedly when a resultant wave function describes the whole state of a physical system, while its projections or single components (*eigenfunctions*) are the result of the particular coordinate system or mathematical space in which the system is described. Giving a reality, therefore, to the mathematical components is somewhat more a matter of convenient choice than physical dictum. We will have much more to say about this when we discuss wave energy, interference effects, and probability amplitudes.

3.2 AN EXPONENTIAL REPRESENTATION

In order to demonstrate a slightly different basis, the foregoing addition of waves will be performed for a sum of exponential rather than sinusoidal functions. Having established the complex function

$$y = ze^{i(\omega t - kx)} = ze^{i\theta} \tag{3.17}$$

where $z = Ae^{i\phi}$, as a solution to the wave equation, we wish to sum two such functions of differing complex amplitude z. First, we note that z, as a complex number of amplitude A and phase ϕ, makes the wave function appear to be

$$y = Ae^{i(\omega t - kx + \phi)} \tag{3.18}$$

In this form the real part of the complex function corresponds precisely to the sinusoidal solution (Eq. 3.1)

$$\text{Re}\, y = A\cos(\omega t - kx + \phi) \tag{3.19}$$

and is the projection of y on the real axis in the complex plane (Fig. 3.3).

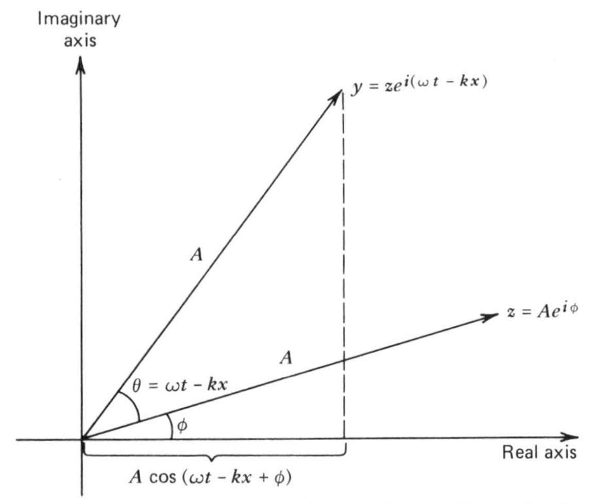

Figure 3.3. Vector representation for a complex function and its projection on real axis.

The complex function y, then, is represented by a rotating vector of magnitude A and a time-dependent phase $(\omega t - kx + \phi)$ in the complex plane. In engineering terminology, this is referred to as the analytic function. Further, the complex conjugate

$$y^* = z^* e^{-i(\omega t - kx)} = Ae^{-i(\omega t - kx + \phi)} \tag{3.20}$$

is also a solution to the wave equation, and represents a complex vector similar to y but rotating in the opposite sense. A sum of the two solutions, by symmetry (Fig. 3.4), must be along the real axis, making

$$y + y^* = 2\,\mathrm{Re}\, y = 2A\cos(\omega t - kx + \phi) \tag{3.21}$$

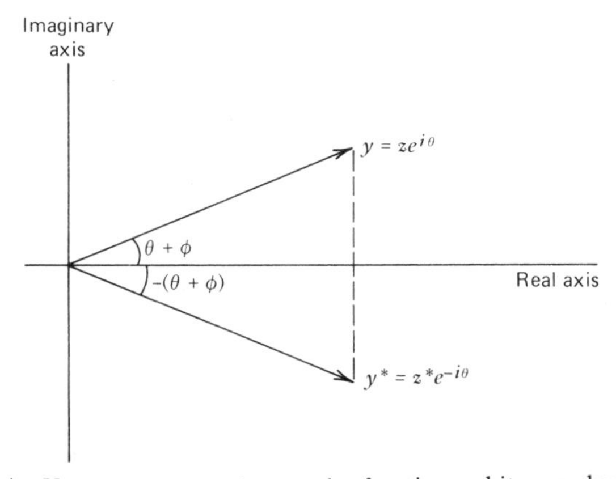

Figure 3.4. Vector symmetry of a complex function and its complex conjugate.

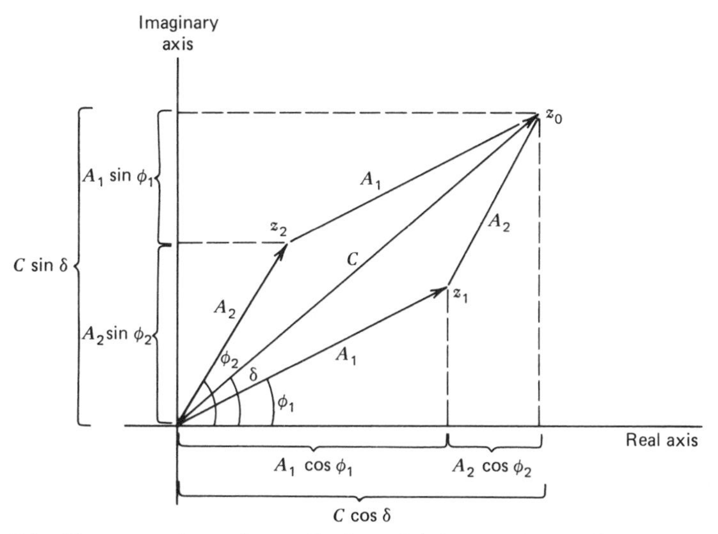

Figure 3.5. Vector sum of complex amplitudes and their projections on the real and imaginary axes.

Returning to the sum of two solutions

$$y = z_1 e^{i\theta} \quad \text{and} \quad y_2 = z_2 e^{i\theta} \tag{3.22}$$

where $z_1 = A_1 e^{i\phi_1}$ and $z_2 = A_2 e^{i\phi_2}$, we attain

$$y = y_1 + y_2 = (z_1 + z_2)e^{i\theta} = z_0 e^{i\theta} \tag{3.23}$$

where $z_0 = Ce^{i\delta}$ is the vector sum of the complex amplitudes z_1 and z_2, shown geometrically in Fig. 3.5. Since the vectors have real magnitudes, A_1, A_2, and C, which form the sides of a triangle, the resultant amplitude C can be gained by forming

$$|z_0|^2 = z_0 z_0^* = (z_1 + z_2)(z_1 + z_2)^*$$

$$= |z_1|^2 + |z_2|^2 + z_1 z_2^* + z_2 z_1^* \tag{3.24}$$

or the more familiar

$$|z_0|^2 = C^2 = A_1^2 + A_2^2 + A_1 A_2 [e^{+i(\phi_1 - \phi_2)} + e^{-i(\phi_1 - \phi_2)}] \tag{3.25}$$

which, using Euler's relation,

$$e^{i(\phi_1 - \phi_2)} + e^{-i(\phi_1 - \phi_2)} = 2\cos(\phi_1 - \phi_2)$$

is again the law of cosines (Eq. 3.10).

Finally, considering that

$$z_0 = z_1 + z_2 \tag{3.26}$$

or

$$Ce^{i\delta} = A_1 e^{i\phi_1} + A_2 e^{i\phi_2}$$

by projecting upon the real and imaginary axes (Fig. 3.5.)

$$C\cos\delta = A_1 \cos\phi_1 + A_2 \cos\phi_2 \tag{3.27}$$

and

$$C\sin\delta = A_1 \sin\phi_1 + A_2 \sin\phi_2$$

we obtain $\tan\phi$ as the same result (Eq. 3.8) for the sum of two sinusoidal functions. Taking the real part of Eq. (3.23), namely,

$$\text{Re}\, y = \text{Re}\, z_0 e^{i(\omega t - kx)} = C\cos(\omega t - kx + \delta) \tag{3.28}$$

we show that exponentials can represent real sinusoidal waves, as long as at the end of any summation of exponentials we use the real part of the function.

3.3 THE ENERGY OF TWO WAVES

The intensity of any single sinusoidal wave is proportional to the square of the amplitude. If we take the amplitude of the sum of two such waves, (differing only in amplitude and phase) to be in the form

$$C^2 = A_1^2 + A_2^2 + 2A_1 A_2 \cos(\phi_1 - \phi_2) \tag{3.29}$$

the average intensity, over any number of complete cycles for the resultant wave to that of its components, is related by

$$I = I_1 + I_2 + 2\sqrt{I_1 I_2}\cos(\phi_1 - \phi_2) \tag{3.30}$$

Thus, the intensity of two superimposed waves differs from the superposition of the individual intensities by the interference term $2\sqrt{I_1 I_2}\cos(\phi_1 - \phi_2)$, where the $\cos(\phi_1 - \phi_2)$ may have any value between $+1$ and -1 pending the phase difference between the two component waves. Only when

$\phi_1 - \phi_2 = \pi/2$ do the waves appear independent of one another, in the sense that the individual intensities simply add, that is, $\cos(\pi/2) = 0$ and $I = I_1 + I_2$. For any other phase difference the intensity of the composite wave could be more or less than the sum of the individual intensities. In particular, if $A_1 = A_2$, making the component waves of equal amplitude, the composite intensity is

$$I = 2I_1[1 + \cos(\phi_1 - \phi_2)] \tag{3.31}$$

which predicts that for two identical waves in phase, $\phi_1 - \phi_2 = 0$, the intensity is $4I_1$. If the waves are perfectly out of phase, $\phi_1 - \phi_2 = \pi$, the intensity is zero. If we represent the extreme cases of interference by the cosine term, we are provoked to understand the appearance or disappearance of the wave energy.

It would be incorrect to say that perfect sinusoidal waves do not exist since we can generate them quite closely. We could argue that waves cannot be independently superimposed on the same medium as we have been doing, in which case we would conclude that there is only the composite wave requiring its appropriate energy. In the above example we would say there is only a wave of amplitude $2A$ requiring an energy related to $4A^2$, and that the two in-phase components of amplitude A do not exist. Or, in the out-of-phase case, we could protest that it takes no energy to produce no wave; hence, nothing exists. This reasoning is compelling since the sinusoidal function itself is a mathematically convenient but arbitrary solution to the wave equation on which to provide a basis for building waves. However, this course of arguing would destroy the principle of superposition which we have been tendering. To preserve the principle, we conclude that the energy must lie elsewhere, and indeed it does.

Even in this somewhat artificial example of two waves on a string, we can propose an exchange of wave energy with the potential energy of the medium itself. The string was stretched under tension before we even engaged it with wave energy. We also demanded small amplitude disturbances and negligible stretch of the string to support the derivation of the wave equation. Thus, the two superimposed waves could borrow or deposit the required amount of energy with the medium to produce the composite wave of amplitude $2A$ or zero. Conversely, if we start with the composite wave energies of the order $4A^2$ and 0, an extraction of the component waves of amplitude A must require the appropriate exchange of energy with the medium. In this way we preserve the principle of superposition and the conservation of energy.

More striking and realistic is the situation where two identical waves are directed to the same point through two different paths (strings), as in Fig. 3.6. If we take the intensity of either wave to be I, the waves should pass

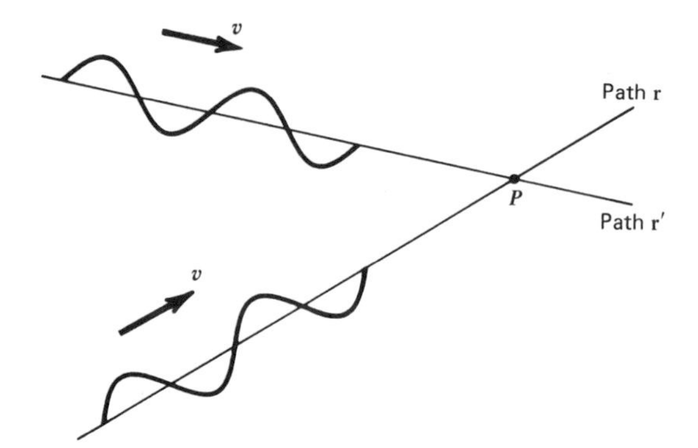

Figure 3.6. Two plane waves crossing at point P.

through each other undisturbed, maintaining their independent paths and intensities. If we place an absorber or otherwise force the waves to combine at point P, the principle of superposition predicts an average or instantaneous intensity between zero and four times that of each individual wave depending upon the phase of the waves. It is assumed that the waves arrive at the point P with their instantaneous displacements along the same line. The prediction is correct. Suppose a knot is tied in the two strings at the point P. Two perfectly out-of-phase waves arriving at the knot yield zero wave intensity as predicted by superposition, but in actuality are depositing energy in the amount of $2I$ capable, for example, of ripping the knot asunder.

Apropos of this intersection of waves, we will discuss the distribution of wave energy as experienced in Thomas Young's interference experiment in which two spherical light waves superimpose upon a screen. An interference pattern, accommodating intensities from zero to $4I$, will emerge as a redistribution of the original wave energy satisfying the conservation of energy. This yet-to-come analysis of Young's experiment was the original impetus for the wave interpretation of light. Finally, in the sense in which we have been discussing superposition, waves do not interfere with one another but merely add point-by-point values of their displacements over space and time, which however confusing, will nonetheless be referred to as interference of waves.

3.4 INTERFERENCE OF TWO OPPOSITELY TRAVELING WAVES

We have previously argued that wave solutions must be in the form $f(t \mp x/v)$, where the minus and plus signs are associated with the direction of travel. Whereas $y = A\cos(\omega t - kx)$ represents a simple plane wave

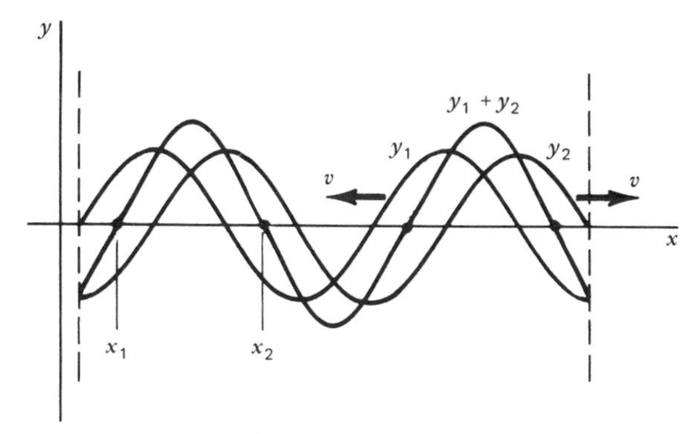

Figure 3.7. Two oppositely traveling waves and their sum. (Note points of zero displacement.)

traveling to the right along the positive x axis, the solution $y = A\cos(\omega t + kx)$ is the same wave propagating in the opposite direction. Since both waves exist simultaneously on the x axis (Fig. 3.7) passing through one another, they can be combined by superposition into the composite function

$$y = y_1 + y_2 = A\cos(\omega t - kx) + A\cos(\omega t + kx) \tag{3.32}$$

Rather than carry out the trigonometric addition indicated, we will perform the summation using the equivalent but more elegant exponential rendition. If

$$y_1 = Ae^{i(\omega t - kx)} = Ae^{i\omega t}e^{-ikx} \tag{3.33}$$

and

$$y_2 = Ae^{i(\omega t + kx)} = Ae^{i\omega t}e^{ikx}$$

represent the two oppositely moving waves, their sum is

$$\begin{aligned} y = y_1 + y_2 &= Ae^{i\omega t}(e^{ikx} + e^{-ikx}) \\ &= 2Ae^{i\omega t}\cos kx \end{aligned} \tag{3.34}$$

the real part of which is the trigonometric sum (3.32), namely,

$$\text{Re}\, y = 2A\cos kx \cos \omega t \tag{3.35}$$

This is interpretable as a position-dependent amplitude $2A\cos kx$, every point of which oscillates in SHM according to the function $\cos \omega t$. Or, if we

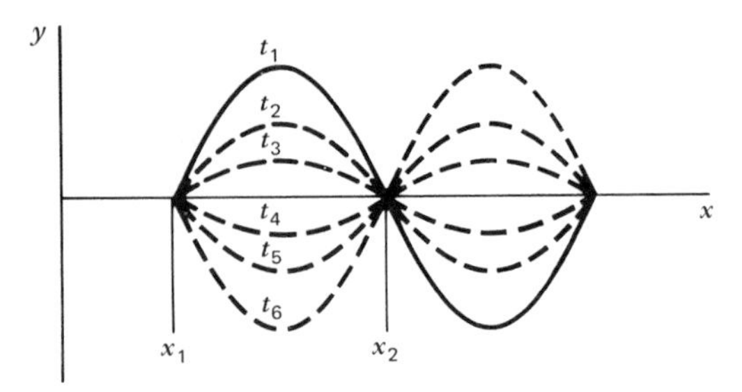

Figure 3.8. A standing wave shown at successive time intervals.

wish, the sum, Eq. (3.35), can be seen as a set of oscillators having a time-dependent amplitude $2A \cos \omega t$ whose displacement in y is modified by the function $\cos kx$. In either view, the sum of opposite traveling waves produces a stationary wave form changing only in size, and having fixed points of zero displacement (Fig. 3.8), determined by the condition

$$\cos kx = 0 \tag{3.36a}$$

making

$$kx_n = (2n + 1)\frac{\pi}{2} \qquad n = 0, 1, 2, \cdots \tag{3.36b}$$

or

$$x_n = \frac{(2n + 1)\pi}{2k} = (2n + 1)\frac{\lambda}{4} \tag{3.36c}$$

These permanent zero displacements or nodal points occur at half wave-length intervals giving credence to the name *standing wave*. In a sense there are no longer any traveling waves, but only the standing amplitude varying wave form.

If we take the distance between two successive nodes to be L, condition (3.36) leads to

$$kL = kx_2 - kx_1 = \pi \tag{3.37}$$

making

$$k = \frac{\pi}{L} \quad \text{or} \quad \lambda = 2L$$

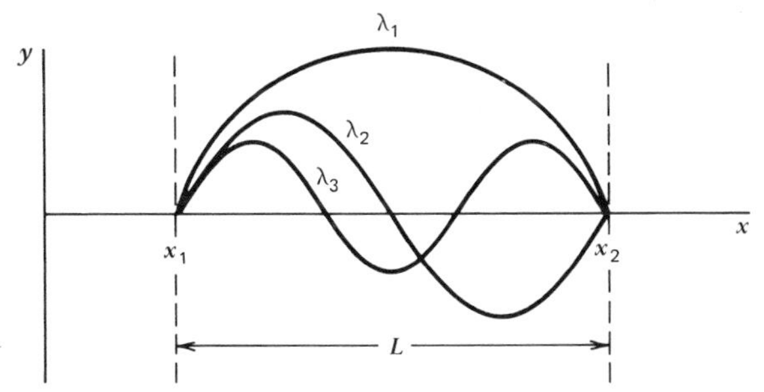

Figure 3.9. The first three allowed harmonics having nodes at x_1 and x_2.

the appropriate wavelength that can be accommodated in the distance L. If we relax the demand for successive nodes, to allow any number of nodes to exist within L, as long as there remains a nodal point at x_1 and x_2, then

$$k_n L = kx_2 - kx_1 = n\pi \tag{3.38}$$

where n is any whole number. The allowed wavelengths having fixed zero displacement at x_1 and x_2, a distance L apart (Fig. 3.9), become $\lambda_n = 2L/n$. Thus, an entire set of standing waves can be accommodated within the interval L, yielding a set of arithmetically decreasing wavelengths and multiple, harmonic frequencies

$$f_n = \frac{v}{\lambda_n} = nf_1 \tag{3.39}$$

where the fundamental frequency is $f_1 = v/2L$, v being the phase velocity of the component traveling waves.

The points of maximum displacement, called *loops* or *antinodes*, occur when $\cos kx = 1$, making their positions located at

$$x_n = \frac{n\pi}{k} = n\frac{\lambda}{2} \qquad n = 0, 1, 2, \cdots \tag{3.40}$$

which, in comparison with Eq. (3.36c), is midway between each node. At these points, (3.40), the energy density is greatest, diminishing to zero at the nodes, and, although the energy varies with position within the standing wave, there is no net flow of energy in either direction.

If, for some reason, walls, mirrors, or the necessary reflecting surfaces should exist at x_1 and x_2 to contain the waves, the most general disturbance within the interval would produce standing wave forms of the type

$$y_n = A_n \cos k_n x \tag{3.41}$$

where $k_n = n\pi/L$, with $n = 1, 2, 3, \cdots$. They would be representative of all the possible harmonics that could survive within the boundary conditions of the interval. Of course, each standing wave component is modified by its time function $\cos \omega t$, but each harmonic could be present and coexist as the resonant evidence of continuous reflection and addition of its traveling wave counterparts.

There is no intention here to pursue standing wave properties except to mention two points of interest. One is that while standing waves are not difficult to produce for mechanical waves, they become quite difficult to observe among light waves where the node-to-node distances are of the order of 10^{-5} cm.[‡] The other point is that even without reflecting walls, the interval L can be looked upon as a cell or modality (*cavity*) in space, within which we can discuss the possible modes or harmonics coexisting in much the same way as for standing waves in a one-dimensional box. Generalized to three dimensions, this aspect of modes will have bearing on the statistical approaches of classical wave theory to the emission and absorption of light.[§] For now, however, we will consider a generalization of one-dimensional standing wave forms leading to Fourier series and Fourier integrals.

3.5 SUM AND SUBSTANCE OF FOURIER SERIES

Combinations of the allowed harmonics are useful in forming more complicated wave forms. Choosing any interval of $2L$ along the x axis guarantees that the time-independent amplitudes of all the allowed standing waves

$$f_n(x) = A_n \cos(k_n x + \phi_n) \tag{3.42}$$

where $k_n = n\pi/L$ must be periodic (Fig. 3.10). The sum of all possible harmonics, obeying the boundary condition

$$f_n(x) = f_n(x + 2L) \tag{3.43a}$$

becomes

$$F(x) = \sum_{n=0}^{\infty} A_n \cos(k_n x + \phi_n) \tag{3.43b}$$

[‡]Ditchburn, R. W. *Light*, Interscience, New York, 1955, Sections 3.22–3 and 14.12.

[§]Loudon, Rodney. *The Quantum Theory of Light*, Oxford Clarendon Press, London, 1973, p. 4.

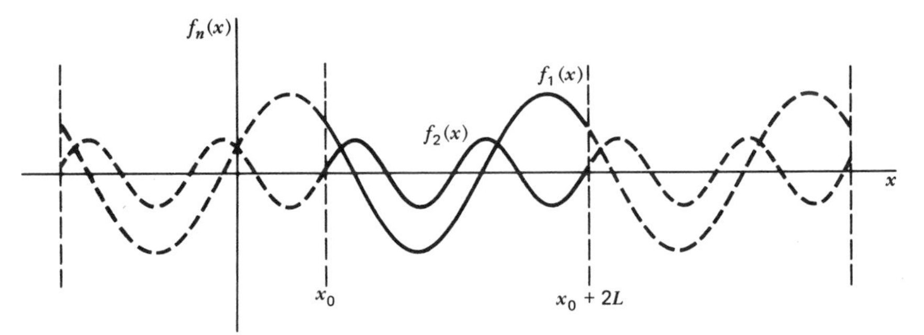

Figure 3.10. A plot of the first two harmonics showing periodicity in the interval $2L$.

which upon expansion yields

$$F(x) = a_0 + \sum_{n=1}^{\infty} (a_n \cos k_n x + b_n \sin k_n x) \tag{3.44}$$

where

$$a_0 = A_0 \cos \phi_0$$

$$a_n = A_n \cos \phi_n$$

$$b_n = -A_n \sin \phi_n$$

The above sum [Eq. (3.44)] is referred to as the Fourier series. It was actually conceived to decompose periodic functions $F(x)$ into their harmonic sinusoidal components, rather than synthesize the harmonics into the composite function we have presented here. In any case, it is Fourier's theorem that reasonably behaved functions $F(\theta)$ (even ones having certain kinds of discontinuities) which are periodic in a range of θ to $\theta + 2\pi$ can be represented by an infinite sum of harmonics as

$$F(\theta) = a_0 + \sum_{n=1}^{\infty} (a_n \cos n\theta + b_n \sin n\theta) \tag{3.45}$$

This expansion is valid for an extraordinarily large class of functions, and a rigorous proof of the theorem with its limits of application is important to its formal use. Here we will only appeal to its plausibility as a method of adding waves and finding their coefficients, in order to extend the notion to the Fourier integral with which we can describe wave packets necessary for building a model of light.[‡]

The essence of finding the coefficients for a particular function rests with the fact that different harmonic functions of the type sin $n\theta$ and cos $n\theta$ are all orthogonal to one another in the following sense: Orthogonality for vectors occurs when the dot product between the two vectors is zero, that is,

$$\mathbf{A} \cdot \mathbf{B} = 0 \tag{3.46}$$

making the vectors at right angles to one another. Orthogonality for the harmonic functions occurs in that

$$\int_{\theta}^{\theta+2\pi} \sin n\theta \sin m\theta \, d\theta = \pi\delta_{m,n}$$

$$\int_{\theta}^{\theta+2\pi} \sin n\theta \cos m\theta \, d\theta = 0$$

and

$$\int_{\theta}^{\theta+2\pi} \cos n\theta \cos m\theta \, d\theta = \pi\delta_{m,n} \tag{3.47}$$

where the Kronecker delta $\delta_{m,n}$ is unity for $n = m$ and zero otherwise. In other words, the product of the different harmonic functions integrated over the repeating interval 2π is zero, an analogue of sorts to orthogonal vectors. Using this property, we can perform the operations

$$\int_{\theta}^{\theta+2\pi} F(\theta)\, d\theta = \int_{\theta}^{\theta+2\pi} a_0 \, d\theta + \int_{\theta}^{\theta+2\pi} (a_n \cos n\theta + b_n \sin n\theta)\, d\theta = 2\pi a_0$$

$$\begin{aligned}\int_{\theta}^{\theta+2\pi} F(\theta)\cos n\theta \, d\theta &= \int_{\theta}^{\theta+2\pi} a_0 \cos n\theta \, d\theta \\ &\quad + \sum_{n=1}^{\infty} \int_{\theta}^{\theta+2\pi} (a_n \cos n\theta + b_n \sin n\theta) \cdot \cos n\theta \, d\theta \\ &= a_n \int_{\theta}^{\theta+2\pi} \cos^2 n\theta \, d\theta = \pi a_n\end{aligned} \tag{3.48}$$

[‡]For formal treatment, see W. Kaplan, *Advanced Calculus*, Addison Wesley, Reading, Mass., 1965, Chapter 7.

and similarly

$$\int_{\theta}^{\theta+2\pi} F(\theta)\sin n\theta \, d\theta = b_n \int_{\theta}^{\theta+2\pi} \sin^2 n\theta \, d\theta = \pi b_n$$

allowing, therefore, an explicit method for finding the coefficients a_0, a_n, and b_n once the particular function $F(\theta)$ is given.

Putting it all together and writing θ as $kx = \pi x/L$, a time-independent periodic wave function $F(x)$ can be expressed as the infinite sum of harmonics

$$F(x) = a_0 + \sum_{n=1}^{\infty} \left(a_n \cos\frac{n\pi x}{L} + b_n \sin\frac{n\pi x}{L} \right) \tag{3.49}$$

where

$$a_0 = \frac{1}{2L} \int_{x}^{x+2L} F(x)\, dx$$

$$a_n = \frac{1}{L} \int_{x}^{x+2L} F(x) \cos\frac{n\pi x}{L}\, dx$$

and

$$b_n = \frac{1}{L} \int_{x}^{x+2L} F(x) \sin\frac{n\pi x}{L}\, dx$$

are the coefficients of the component harmonics that make up the stationary wave function $F(x)$. Whether we consider the components as real or not is a moot question, the reality being one of mathematical convenience or experimental arrangement as we will see when we consider the nature of light.

The energy associated with a wave of the type $F(x)$ is related to the square of the function. Carrying out a calculation for the energy in one periodic interval

$$\int_{x}^{x+2L} [F(x)]^2 dx = \int_{x}^{x+2L} \left[a_0 + \sum_{n=1}^{\infty} \left(a_n \cos\frac{n\pi x}{L} + b_n \sin\frac{n\pi x}{L} \right) \right]^2 dx \tag{3.50}$$

which upon expanding the square and eliminating orthogonal terms leaves the energy related to

$$\text{Energy} \sim a_0^2(2L) + \sum_{n=1}^{\infty} \left(a_n^2 + b_n^2\right)L \tag{3.51}$$

This is the Fourier energy theorem which shows the total energy to be the sum of energies for each Fourier component.

3.6 PASSING TO THE FOURIER INTEGRAL

As an extension to the method of Fourier series, the finite interval $2L$ which we have been using for periodic functions can be extended in size to infinity. This will lead to an integral theorem allowing a synthesis or analysis of waves for a continuous distribution of wavelengths or frequencies, not just harmonic multiples.

Thus far, we have entertained the Fourier series as a sum of stationary sinusoidal waves of multiple wave numbers $k_n = nk_1$, where the base value k_1 is π/L. Expressing the sinusoidal terms in their equivalent exponential form

$$\cos nk_1x = \frac{e^{ink_1x} + e^{-ink_1x}}{2} \tag{3.52}$$

and

$$\sin nk_1x = i\frac{\left(e^{-ink_1x} - e^{ink_1x}\right)}{2}$$

we can express the Fourier series (3.49) as

$$F(x) = \sum_{n=-\infty}^{\infty} c_n e^{ink_1x} \tag{3.53}$$

where

$$c_n = \frac{a_n - ib_n}{2}$$

$$c_{-n} = \frac{a_n + ib_n}{2}$$

and

$$c_0 = a_0$$

The new complex amplitudes are all obtained from

$$c_n = \frac{1}{2L}\int_x^{x+2L} F(x)e^{-ink_1x}dx \tag{3.54}$$

since e^{-ink_1x} and e^{imk_1x} are orthogonal for $n \neq m$, and unity for $n = m$.

So far, this is none other than the original Fourier series expressed in exponential form. The negative n poses no problem since, although there is no physical counterpart, the negative and positive exponential harmonics are always combined in pairs to construct the real functions that describe real waves having harmonic numbers of positive n only.

It is now, however, by extending the interval $2L$ to infinity that we gain a new physical sense for the function $F(x)$ and its expansion in what will become a continuum of k values. For convenience we will set x to be $-L$, making the expansion of the periodic function

$$F(x) = \frac{1}{2L}\sum_{n=-\infty}^{\infty} e^{ink_1x}\int_{-L}^{L} F(x')e^{-ink_1x'}dx' \tag{3.55}$$

where the dummy variable under the integral has been labeled x'. Allowing the interval $2L$ to grow without bound and inserting $k_1 = \pi/L$, we have

$$F(x) = \lim_{L\to\infty}\frac{1}{2\pi}\sum_{n=-\infty}^{\infty} k_1\int_{-L}^{L} F(x')e^{-ink_1(x'-x)}dx' \tag{3.56}$$

which can be evaluated by appeal to the limiting process. As L becomes large, k_1 becomes small, and so accordingly does the interval Δk diminish.

$$\Delta k = k_{n+1} - k_n = \frac{\pi}{L} = k_1 \tag{3.57}$$

Since $nk_1 = k_n$, in the limit as $k_1 = \Delta k$ tends toward zero, the set of k_n approaches the refinement of a continuum.

$$\lim_{L\to\infty} nk_1 = \lim_{L\to\infty} k_n = k \tag{3.58}$$

Looking upon the infinite sum from Eq. (3.56) as

$$\sum_{n=-\infty}^{\infty} \Delta k G(k_n) \tag{3.59}$$

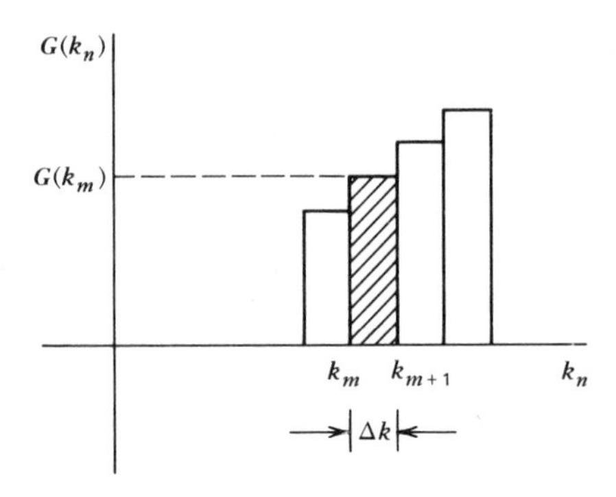

Figure 3.11. Rectangular area representing each element of an infinite sum.

we find that each term is the area of a small rectangle of width Δk (Fig. 3.11), which in the limit of Δk approaching zero and $G(k_n)$ becoming $G(k)$ is the definition of a definite integral. Equation (3.56) written as

$$F(x) = \lim_{L\to\infty} \frac{1}{2\pi} \sum_{k=-\infty}^{\infty} \Delta k \int_{-L}^{L} F(x')e^{-ik_n(x'-x)}dx' \qquad (3.60)$$

finally becomes

$$F(x) = \frac{1}{2\pi}\int_{-\infty}^{\infty} dk \int_{-\infty}^{\infty} F(x')e^{ik(x-x')}dx' \qquad (3.61)$$

If we define $G(k)$, remembering that x' is a dummy variable, by

$$G(k) = \frac{1}{\sqrt{2\pi}}\int_{-\infty}^{\infty} F(x)e^{-ikx}dx \qquad (3.62a)$$

then

$$F(x) = \frac{1}{\sqrt{2\pi}}\int_{-\infty}^{\infty} G(k)e^{ikx}dk \qquad (3.62b)$$

Equation (3.61) is the Fourier integral and $F(x)$ and $G(k)$ are referred to as the Fourier transforms of one another. It is a powerful tool for wave addition because, given the amplitudes $G(k)$ for a set of plane waves of corresponding wave number k, we can synthesize the group into the composite function $F(x)$, which no longer needs to be a periodic function and henceforth will be called a *wave packet*. But equally true, viewing the symmetry of the integrals, we can decompose any wave packet $F(x)$ into elemental plane waves of amplitude $G(k)$. Either function contains all the

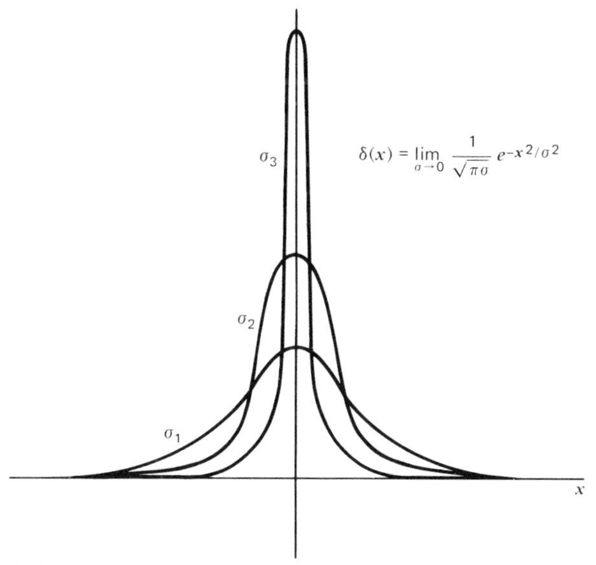

Figure 3.12. Approaching a delta function having infinite height and unit area.

information about the packet, which is nicely illustrated by an energy consideration. The energy contained in a wave packet is

$$\int_{-\infty}^{\infty} [F(x)]^2 dx = \int_{-\infty}^{\infty} [G(k)]^2 dk \tag{3.63}$$

which is Parseval's formula. As such, the energy of a packet is expressible in an x-space or k-space calculation (where k becomes a variable in a wave number coordinate system, just as x is a coordinate variable in a spatial coordinate system).

Returning to the Fourier integral (3.61) we can gain an interesting function called the Dirac delta function. When we define the integration over k for

$$\frac{1}{2\pi} \int_{-\infty}^{\infty} dk e^{ik(x-x')} = \delta(x - x') \tag{3.64}$$

then $\delta(x - x')$, the Dirac delta, must be a function for which

$$\int_{-\infty}^{\infty} F(x')\delta(x - x')\, dx' = F(x) \tag{3.65}$$

that is, the delta function must be zero at all x' except $x' = x$, where it must be infinite in such a way that its integral is unity.

$$\int_{-\infty}^{\infty} \delta(x)\, dx = 1 \tag{3.66}$$

It is not a normal mathematical function but can be approximately envisioned by (Fig. 3.12)

$$\delta(x) = \lim_{\sigma \to 0} \frac{1}{\sqrt{\pi}\sigma} e^{-x^2/\sigma^2} \tag{3.67}$$

3.7 FROM PLANE WAVES TO GAUSSIAN PACKETS

We are now in a position to ask what form the coefficient function $G(k)$ must take for any wave function $F(x)$. Conversely, given the function of the amplitudes $G(k)$, what must be the form of the wave packet $F(x)$. A simple case, with which we are already familiar, is the plane monochromatic wave having a wave number $k = k_0$. With its amplitude $G(k)$ defined as related to the Dirac delta $\delta(k - k_0)$, the Fourier transform for $G(k) = A\delta(k - k_0)$ is

$$F(x) = \frac{A}{\sqrt{2\pi}} \int_{-\infty}^{\infty} \delta(k - k_0) e^{ikx} dk = \frac{A}{\sqrt{2\pi}} e^{ik_0 x} \tag{3.68}$$

the real part of which is the familiar sinusoidal wave (Fig. 3.13).

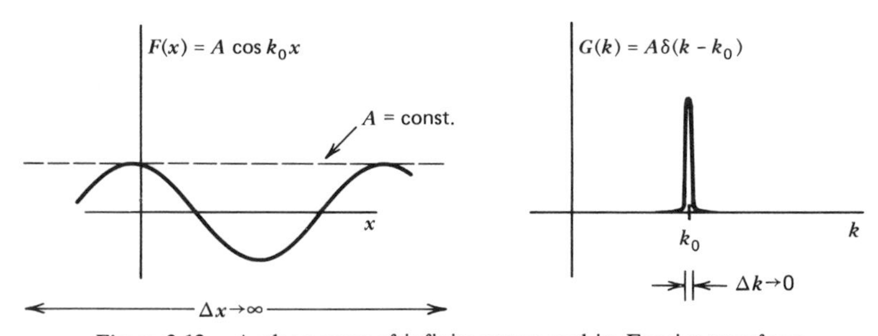

Figure 3.13. A plane wave of infinite extent and its Fourier transform.

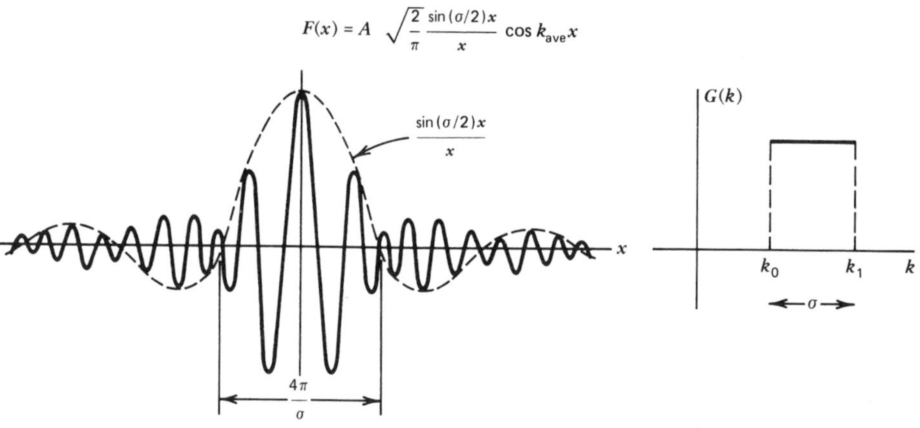

Figure 3.14. A spread of wave numbers $G(k)$ and the corresponding wave packet $F(x)$.

More interesting is the example of a spread in k values, where

$$G(k) = \begin{cases} A & \text{for} \quad k_0 \leqslant k \leqslant k_1 = k_0 + \sigma \\ 0 & \text{for} \quad k_0 > k > k_0 + \sigma \end{cases} \tag{3.69}$$

The Fourier transform of this function is

$$F(x) = \frac{1}{\sqrt{2\pi}}\int_{k_0}^{k_0+\sigma} Ae^{ikx}dk = \frac{A}{\sqrt{2\pi}}\left.\frac{e^{ikx}}{ix}\right|_{k_0}^{k_0+\sigma} \tag{3.70}$$

which can be written as

$$\frac{A}{\sqrt{2\pi}}e^{i(k_0+\sigma/2)x}\left[\frac{e^{i(\sigma/2)x} - e^{-i(\sigma/2)x}}{ix}\right] = A\sqrt{\frac{2}{\pi}}\,e^{i(k_0+\sigma/2)x}\frac{\sin(\sigma/2)x}{x} \tag{3.71}$$

The real part of this result is a plane wave of modified amplitude

$$F(x) = A\sqrt{\frac{2}{\pi}}\,\frac{\sin(\sigma/2)x}{x}\cos k_{\text{ave}}x \tag{3.72}$$

where $k_{\text{ave}} = k_0 + \sigma/2$. The envelope departs radically from a sine curve at small x (Fig. 3.14). The departure, however, is less dramatic as the spread in

k is reduced. With σ approaching zero, $F(x)$ approaches a plane wave and $G(k)$ narrows towards a delta function. On the contrary, as the spread in k values increases, σ becomes very large and the major portion of $F(x)$, which lies between $x = \pm 2\pi/\sigma$, shrinks towards a delta function. This relationship of the corresponding widths of the packets is quite significant not only for classical waves but especially for the uncertainty principle of quantum theory.

If we take the significant part of the wave packet (basically, the central maximum in which most of the energy is concentrated) as approximately of the order $\Delta x \simeq 4\pi/\sigma$, the product of the widths of $G(k)$ and $F(x)$ is of the order

$$\Delta k \Delta x \simeq 1 \tag{3.73}$$

In no case can we further reduce both Δk and Δx simultaneously. Any narrowing of one of the packet widths causes an increase in the other.

To further illustrate this relation of spread in position to spread in wave number consider a limited plane wave

$$F(x) = \begin{cases} Ae^{ik_0 x} & -\dfrac{1}{\sigma} \leqslant x \leqslant \dfrac{1}{\sigma} \\ 0 & |x| > \dfrac{1}{\sigma} \end{cases} \tag{3.74}$$

Finding the packet in k-space from the transform

$$G(k) = \frac{A}{\sqrt{2\pi}} \int_{-1/\sigma}^{1/\sigma} e^{ik_0 x} e^{-ikx} dx = \frac{A}{\sqrt{2\pi}} \left. \frac{e^{i(k_0 - k)x}}{k_0 - k} \right|_{-1/\sigma}^{1/\sigma} \tag{3.75}$$

we have

$$G(k) = A\sqrt{\frac{2}{\pi}} \frac{\sin[(k_0 - k)/\sigma]}{k_0 - k} \tag{3.76}$$

This is just the envelope of the previous example, but in k-space. The plane wave has an envelope similar to the spread in k values of the previous example. The width of the central maximum of the function $G(k)$, Fig. 3.15, is of the order of $2\pi\sigma$, verifying again the relationship of the spread in values of Δx and Δk, Eq. (3.73). As σ approaches zero, the plane wave becomes infinite and $G(k)$ tends toward a delta function.

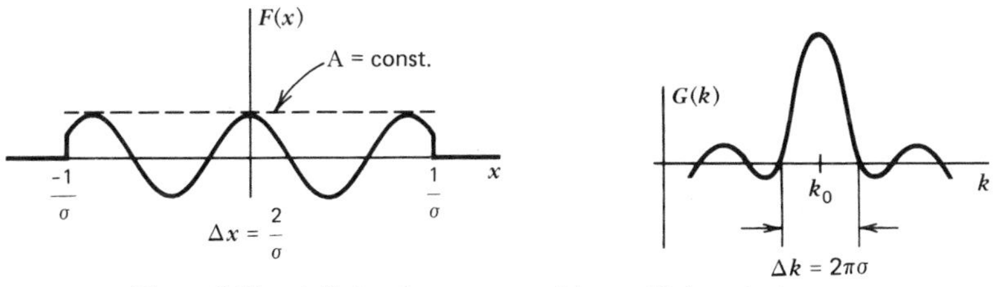

Figure 3.15. A finite plane wave and its coefficients in k-space.

Finally, we seek a function which transforms into itself. This requirement is satisfied with the Gaussian function that modifies a plane wave:

$$F(x) = Ae^{-x^2/2\sigma^2}e^{ik_0x} \tag{3.77}$$

The Fourier transform is

$$G(k) = \frac{A}{\sqrt{2\pi}}\int_{-\infty}^{\infty} e^{-x^2/2\sigma^2}e^{i(k_0-k)x}dx = \frac{A}{\sqrt{2\pi}}\int_{-\infty}^{\infty} e^{-[x^2/2\sigma^2 - i(k_0-k)x]}dx \tag{3.78}$$

By setting $a^2 = 1/2\sigma^2$ and $b = i(k_0 - k)$, and completing the square, we have

$$e^{-(a^2x^2 - bx)} = e^{-[(ax-b/2a)^2 - b^2/4a^2]} = e^{b^2/4a^2}e^{-(ax-b/2a)^2} \tag{3.79}$$

Letting $u = ax - b/2a$ and using

$$\int_{-\infty}^{\infty} e^{-u^2}du = \sqrt{\pi} \tag{3.80}$$

we arrive at

$$G(k) = \frac{A}{\sqrt{2\pi}}\frac{e^{b^2/4a^2}}{a}\int_{-\infty}^{\infty} e^{-u^2}du = A\sigma e^{-(k_0-k)^2\sigma^2/2} \tag{3.81}$$

which is also a Gaussian function (Fig. 3.16). Taking the spread of the wave packets $F(x)$ and $G(k)$ to be measured at those points where the exponentials drop off to the value e^{-1}, their widths become $\Delta x = 2\sqrt{2}\,\sigma$ and $\Delta k =$

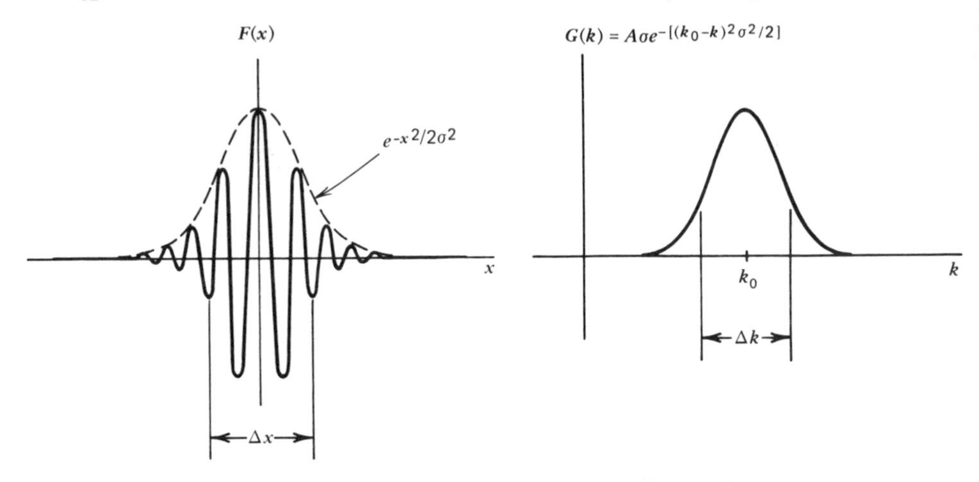

Figure 3.16. The Gaussian wave packet and its transform.

$2\sqrt{2}/\sigma$, their product again being close to but greater than unity. The value $\Delta k \Delta x$ is actually less than the previous examples, and as we shall see in quantum applications, the Gaussian wave functions hold a special position because the product of their widths are the least for any combination of wave packets and their transforms. For this reason the Gaussian is referred to as the *function of least uncertainty*.

By now we can gain physical insight into the properties of wave packets and their transforms by scanning the pairs in each example shown in Fig. 3.13 through Fig. 3.16. In each case, Δx and Δk range inversely to one another. We have gone from a single k value $G(k) = \delta(k - k_0)$ for which $F(x)$ is an infinite plane wave to a spread in k values for which $F(x)$ becomes more localized, its energy being contained in a smaller width Δx. In a very broad spread in k values, $F(x)$ tends toward a delta function with its energy concentrated about a definite point.

This is also nicely seen by considering the Gaussian function for three different values of σ. With Fig. 3.16 representing a moderate value of σ, as σ approaches infinity the spread in k values approaches a delta function, while $F(x)$ stretches out to a long cosine wave (Fig. 3.17*a*). When σ grows very small, $G(k)$ becomes approximately constant over a broad range of k values while the Gaussian wave packet $F(x)$ shrinks towards the delta function (Fig. 3.17*b*). In this latter case, the energy becomes so localized that even classically, the wave $F(x)$ takes on the appearance of a massless particle located within an extremely narrow position Δx.

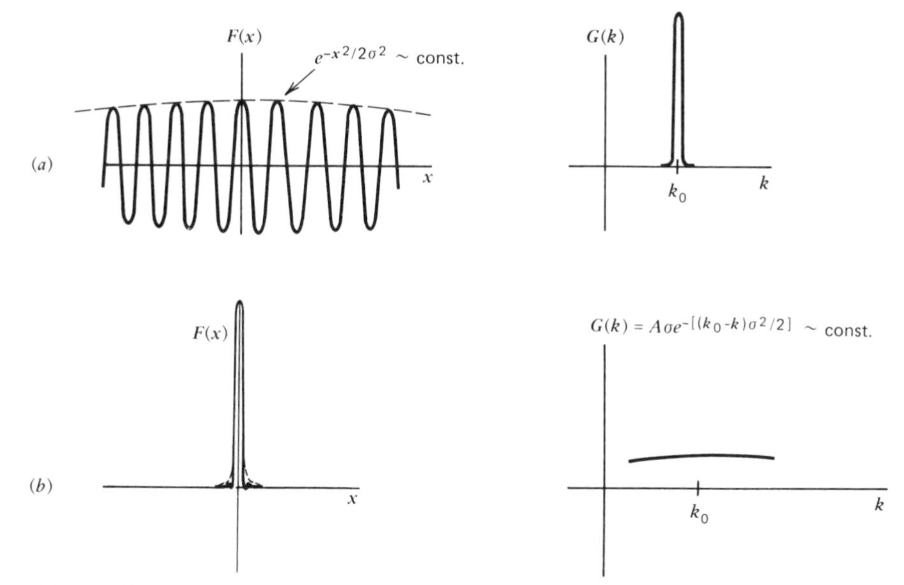

Figure 3.17. The Gaussian function and its transform for (*a*) σ very large; (*b*) σ very small.

3.8 TIME DEPENDENCE AND SUPERPOSITION

At the beginning of Section 3.5, we stated that Fourier series were a sum of harmonics representing the time-independent amplitudes of standing waves. However, these harmonic components can also be considered as the form of a traveling wave at time, $t = 0$. Citing a traveling wave as

$$y = A\cos(kx - \omega t) \tag{3.82}$$

at time $t = 0$, we have

$$y = A\cos kx \tag{3.83}$$

If we choose a set of wave numbers $k_n = n\pi/L$, the harmonic functions (3.83) become the components of the Fourier series. By also considering the time dependence of waves of multiple k values in superposition, we can write the Fourier sum as

$$F(x,t) = \sum_{n=-\infty}^{\infty} A_{k_n} e^{i(k_n x - \omega_n t)} \tag{3.84}$$

where the angular frequency ω is labeled with the subscript n, since in general ω is a function of k_n, the simplest case being one in which each component propagates at the same velocity

$$\omega_n = vk_n \tag{3.85}$$

This simple relation will not always be true, as we shall see shortly, where in a dispersive medium waves of different k value propagate at slightly different velocities.

With the same justification brought to the derivation of the Fourier integral, namely, that the amplitudes become a continuous function of k as the periodic interval approaches infinity

$$\lim_{L\to\infty} A_{k_n} - A_{k_{n+1}} = A(k)\,dk \tag{3.86}$$

Eq. (3.84) can be rewritten as

$$F(x,t) = \frac{1}{\sqrt{2\pi}}\int_{-\infty}^{\infty} A(k)e^{i(kx-\omega t)}dk \tag{3.87}$$

with the corresponding transform becoming

$$A(k)e^{-i\omega t} = \frac{1}{\sqrt{2\pi}}\int_{-\infty}^{\infty} F(x,t)e^{-ikx}dx \tag{3.88a}$$

which, since ω is not a function of x, can be expressed as

$$A(k) = \frac{1}{\sqrt{2\pi}}\int_{-\infty}^{\infty} F(x,t)e^{-i(kx-\omega t)}dx \tag{3.88b}$$

But $A(k)$, which is the same as the $G(k)$ we have been using, is the Fourier transform of $F(x,0)$, making

$$F(x,t) = F(x,0)e^{-i\omega t} \tag{3.89}$$

and the amplitudes, which are time-independent, become

$$A(k) = \frac{1}{\sqrt{2\pi}}\int_{-\infty}^{\infty} F(x,0)e^{-ikx}dx \tag{3.90}$$

Before embarking on the use of this time-dependent expression for the Fourier wave packet, we shall depart first to consider a simple case of the

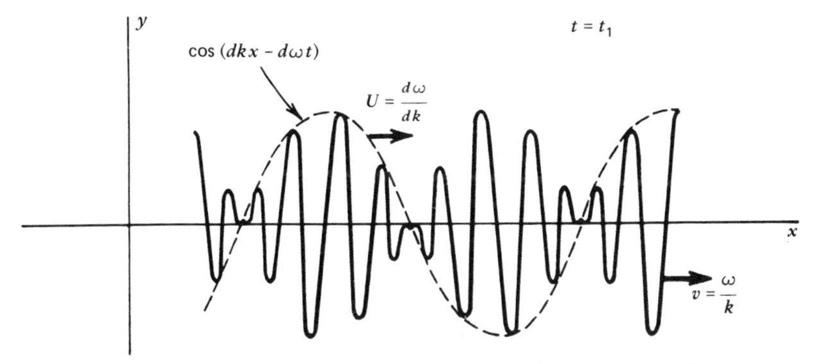

Figure 3.18. The sum of two waves of slightly differing angular frequency and wave number.

addition of two traveling waves having slightly differing k values. Since k is related to both v and ω the implication is that either or both the propagation speed and angular frequency must vary.

With the two traveling waves expressed as

$$y_1 = Ae^{i[(k+dk)x-(\omega+d\omega)t]} \tag{3.91}$$

and

$$y_2 = Ae^{i[(k-dk)x-(\omega-d\omega)t]}$$

the sum becomes

$$y = y_1 + y_2 = Ae^{i(kx-\omega t)}\left[e^{i(dkx-d\omega t)} + e^{-i(dkx-d\omega t)}\right] \tag{3.92}$$

the real part of which is

$$y = 2A\cos(dkx - d\omega t)\cos(kx - \omega t) \tag{3.93}$$

Thus, we have a traveling wave of average angular frequency and wave number, ω and k, moving with a phase velocity $v = \omega/k$. It is modulated, however, by an envelope $\cos\{dk[x - (d\omega/dk)t]\}$ which has a small wave number dk (long wave length) and a speed $U = d\omega/dk$, as pictured at one instant in Fig. 3.18. If ω is proportional to k, as is true if each component moves with the same velocity, the envelope which represents the difference wave also propagates at the same velocity, $U = d\omega/dk = v = \omega/k$. But if each component has a different phase velocity, then $d\omega/dk$ which is referred to as the *group velocity*, can be quite different from the phase velocity. In particular, we can obtain

$$U = \frac{d\omega}{dk} = \frac{d(vk)}{dk} = v + k\frac{dv}{dk} \tag{3.94}$$

If the phase velocity decreases with increasing wave number, which is the case for light in a dispersive medium, dv/dk is negative and the group velocity is less than the phase velocity. The energy carried by the wave would propagate at the group velocity, and the high speed phase waves would begin under the tail of the envelope moving through it only to die out at the head. The situation is similar to the higher speed ripples seen on a caterpillar while its body progresses at a smaller group velocity.

Having considered the time-dependent sum of two waves of slightly different wave number, we can now consider a narrow spread in k values encountered in, for example, a Gaussian distribution, Eq. (3.81)

$$A(k) = \sigma e^{-(k-k_0)^2\sigma^2/2} \tag{3.95}$$

The wave packet is represented by

$$F(x,t) = \frac{\sigma}{\sqrt{2\pi}}\int_{-\infty}^{\infty} e^{-(k-k_0)^2\sigma^2/2} e^{i(kx-\omega t)}dk \tag{3.96}$$

Assuming ω to be a slowly varying function of k in a narrow range of wave numbers, we can expand ω in a Taylor series as

$$\omega(k) = \omega_0 + \left(\frac{d\omega}{dk}\right)_{k_0}(k-k_0) + \frac{1}{2}\frac{d^2\omega}{dk^2}(k-k_0)^2 + \cdots \tag{3.97}$$

If we use the first two terms and set $U = (d\omega/dk)_{k_0}$, the integral becomes

$$F(x,t) = \frac{\sigma}{\sqrt{2\pi}} e^{i(k_0x-\omega_0 t)}\int_{-\infty}^{\infty} e^{-[(k-k_0)^2\sigma^2/2 - i(k-k_0)(x-Ut)]}dk \tag{3.98}$$

But this is in the same form as the integral (3.78) where letting $a^2 = \sigma^2/2$ and $b = -i(x-Ut)$, Eq. (3.98) becomes

$$F(x,t) = \frac{\sigma}{\sqrt{2\pi}} e^{i(k_0x-\omega_0 t)}\frac{e^{b^2/4a^2}}{a}\int_{-\infty}^{\infty} e^{-u^2}du$$

$$= e^{i(k_0x-\omega_0 t)}e^{-(x-Ut)^2/2\sigma^2} \tag{3.99}$$

which is a plane traveling wave $e^{i(k_0x-\omega_0t)}$ having the central wave number and angular frequency of the narrow distribution of k values, moving under a Gaussian envelope the center of which is progressing with the group velocity $U = (d\omega/dk)_{k_0}$. At any given instant $F(x, t)$ and $A(k)$ resemble the time-independent transforms (3.77) and (3.81), shown in Fig. 3.16.

An interesting property of this Gaussian-shaped traveling wave is that when the Gaussian has the value e^{-1}, its width is

$$\Delta x = 2(\sqrt{2}\,\sigma + Ut) \tag{3.100}$$

which is increasing with time. Thus, a Gaussian packet, having a reasonably narrow spread of wave numbers, will spread in time as it propagates, approaching the form of a plane wave. Only the center of the packet is moving at the group velocity U, other parts move faster and slower than U to promote the spread. The approximations (3.99) and (3.100) become poor for large t and small σ since higher order terms in the expansion for ω become important, but the results indicate that a Gaussian distribution reasonably localized in time and space will spread rapidly at first, slowing down in the spread as time progresses and the packet broadens into a plane wave.

REFERENCES

Ditchburn, R. W. *Light*, Interscience, New York, 1955.

Jenkins, F. A. and H. E. White. *Fundamentals of Optics*, McGraw-Hill, New York, 1950.

Kaplan, W. *Advanced Calculus*, Addison-Wesley, Reading, Mass., 1965.

Loudon, R. *The Quantum Theory of Light*, Oxford Clarendon Press, London, 1973.

Powell, J. L. and B. Crasemann. *Quantum Mechanics*, Addison-Wesley, Reading, Mass., 1965.

PROBLEMS

3.1. Carry out the addition of three sinusoidal traveling waves differing only in their phase by $|\Delta\phi| = 2\pi/3$.

3.2. Repeat the procedure of Problem 3.1 but use the exponential representation for the traveling waves.

3.3. Find the energy of the three waves of Problem 3.1.

3.4. When two waves moving along different paths cross at a point, how do the waves add at that point? What can you say about the planes of vibration? A careful analysis requires a discussion of polarization.

3.5. If the two waves of the previous problem arrive at a point on a screen (or a fixed knot in the case of mechanical waves) completely out of phase by π, where does the energy contained in the waves go?

3.6. Explain the energy distribution in standing waves, especially when the pattern represents maximum displacements and zero displacement. Derive the results in Eqs. (3.47).

3.7. Solve for the coefficients in the Fourier series expansion of the function $F(x) = +c$ (a constant) for $0 < x < L$ and $F(x) = -c$ for $L < x < 2L$. This is a square wave.

3.8. Compute the energy associated with a square wave.

3.9. Using the Fourier transform, find $G(k)$ for the function $F(x) = c$ (a constant) in the interval $-1/\sigma \leqslant x \leqslant 1/\sigma$.

3.10. Plot the functions in Problem 3.9 and evaluate $\Delta x \Delta k$.

3.11. In deriving the time-dependent Fourier transform for the Gaussian distribution (3.95), find the function $F(x, t)$ using the third term in the Taylor series (3.97) calling $W = \frac{1}{2} d^2\omega/dk^2$. Show that $F(x, t)$ still has a Gaussian profile and that its width is increasing with time according to $\Delta x = \Delta x_0 \sqrt{1 + t^2/T^2}$ where T is a constant.

3.12. The Dirac delta function is sometimes represented by

$$\delta(x) = \lim_{\sigma \to 0} \frac{1}{\pi} \frac{\sigma}{x^2 + \sigma^2}$$

Show that this exhibits the same properties as represented by Eq. (3.67).

3.13. Derive the value of the product $\Delta k \Delta x$ for the Gaussian wave function and its transform. Show that this product is independent of σ, and smaller than for any other wave form you might choose.

3.14. Show that group velocity can be expressed by $U = -\lambda^2 df/d\lambda$.

4

Light as a Wave

Light is radiant energy. It is created in atomic processes, emitted by sources, propagated through space, and absorbed by matter. A wave model for the nature of light was compellingly introduced by Thomas Young in 1801. This model predicted behavior unaccounted for by the Newtonian corpuscular concept. Within the same century, Maxwell's equations established the electromagnetic basis for light.

Starting with idealized simple waves we will gain an understanding of interference phenomena and their attendant energy distributions. A closer look at real light, however, will reveal the more complicated features of coherence and correlation in which the time-averaged intensity, rather than the wave displacements, becomes the best observable feature. In a consideration of both sources and detection of light, the limitations of the wave theory emerge.

4.1 A YOUNG WAVE SOLUTION

Having considered some of the wave solutions and their properties we are now interested in understanding their applications to light. If light can be shown to superimpose and interfere, characteristics unique to the classical wave, then the wave nature of light can be established. Thomas Young devised just such an experiment.

By allowing sunlight to pass through a pinhole in an otherwise opaque screen S_1, and interposing a second screen S_2 containing two pinholes, as in Fig. 4.1, Young achieved an interference pattern of lighter and darker bands on the final screen S_3 (Fig. 4.2). Not only did this result defy explanation by the then held Newtonian particle concept, but it prompted Young to introduce a wave model for light.

Under the Newtonian view, the original light from the sun consisted of particles entering on the x axis through the first pinhole. In order for some light particles to reach the second set of pinholes on S_2, the Newtonian

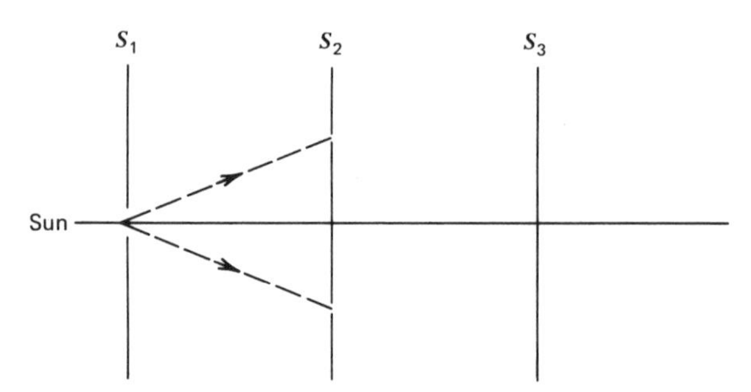

Figure 4.1. Opaque screens with pinholes for Young's experiment.

corpuscles would have to be splayed in all directions by some interaction with the material around the pinhole (Fig. 4.3). Here we have a set of possibilities: either the particles continue through the second set of pinholes on a straight line making two spots of light on S_3, or again by some interaction within the pinholes they become dispersed in many or all directions toward the final screen.

Young's fringes show a definite distribution of the light energy with the brightest fringe centered on the x axis. The Newtonian particles would have either to annihilate each other at certain locations or arrive only at predetermined positions in order to reproduce the interference pattern. Since the Newtonian model could not provide the mechanism for this distribution of energy, it failed to explain Young's experiment. Interestingly enough, although we are condemning the Newtonian corpuscle, we will return to this same analysis, but with the massless particles (photons) of quantum theory, to exact an explanation of the fringe pattern.

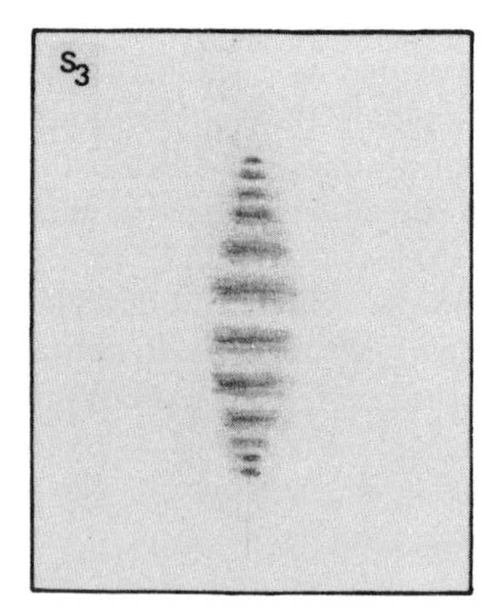

Figure 4.2. Young's interference bands.

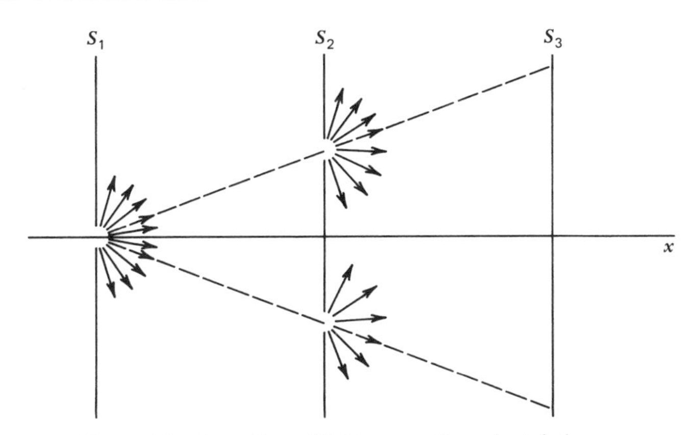

Figure 4.3. Spreading of light energy through pinholes.

In the meantime, Young found that by assigning a wave nature to the light he could account for the pattern in terms of an interference of waves. Of course, Young had to make some simplifying assumptions to achieve this interpretation. A closer look at some of the assumptions will occupy a good part of this chapter and make it appear astounding that Young ever produced a fringe pattern in the first place.

To begin, we shall assume that an infinite continuous monochromatic plane wave

$$f(x, t) = A_0 \cos(kx - \omega t) \tag{4.1}$$

is traveling along the x axis impinging upon the first pinhole. If the portion of this wave front encountering the small area of the pinhole merely traveled straight through along the x axis, it would never reach the second set of pinholes but instead be reflected or absorbed at the screen S_2. The assumption must be made that the light emerges from the first pinhole, spreading in all directions. This assumption is at the heart of Huygens' principle which holds that every point on a wave front acts as a point source emitting secondary wavelets in all directions. Although the principle is somewhat crude, it provides a mechanism for constructing advancing wave fronts, and in the case of an idealized infinitesimal pinhole, it predicts a spherical wave expanding like a bubble toward the second set of pinholes (Fig. 4.4).

In a more realistic setting the pinholes have a small but finite area for which a diffraction calculation would have to be made to determine the precise wave form arriving at the secondary pinholes. In either case the waves impinging upon the second set of pinholes can still be assumed to be

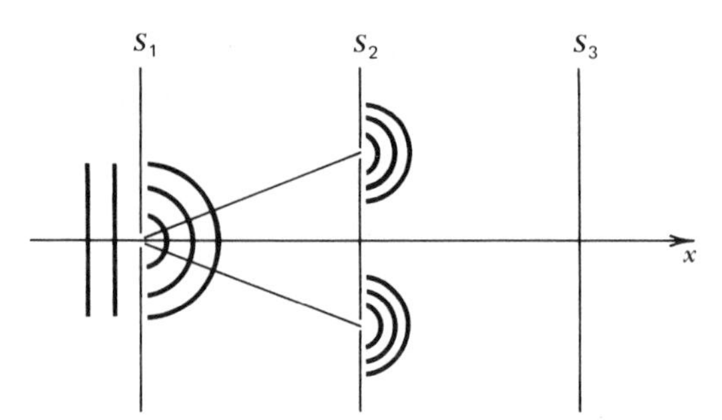

Figure 4.4. Wave fronts emerging from pinholes.

approximately sinusoidal plane waves having some fixed relative phase difference to one another since they both came from the same, approximately infinitesimal, source. If the geometry is fixed so that the pinholes are equidistant from the source, the relative phase difference can be taken as zero.

Again, for the same arguments, two expanding wave fronts will emerge from the second set of pinholes, each acting as new point sources. Thus, the light energy encountered on the third screen at any point $P(\mathbf{r})$ is the combination of two spherical waves having the form

$$f(\mathbf{r}, t) = \frac{uA}{|\mathbf{r}|}\cos(\mathbf{k} \cdot \mathbf{r} - \omega t + \phi) \tag{4.2}$$

where u is a function of the geometry representing the result of the diffraction calculation at the pinhole. More commonly, rectangular slits are used in Young's experiment oriented along the z axis (out of the page). They are easier to construct and control, and allow the passage of more light energy.

Taking an origin midway between the two pinholes which are a distance d apart and letting $\mathbf{r}_1$, $\mathbf{r}_2$, and $\mathbf{r}$ represent radial vectors to each pinhole and the point on the screen (Fig. 4.5), we can express the two waves as

$$f_1 = \frac{u_1 A_1}{|\mathbf{r} - \mathbf{r}_1|}\cos(k|\mathbf{r} - \mathbf{r}_1| - \omega t + \phi_1) \tag{4.3}$$

and

$$f_2 = \frac{u_2 A_2}{|\mathbf{r} - \mathbf{r}_2|}\cos(k|\mathbf{r} - \mathbf{r}_2| - \omega t + \phi_2)$$

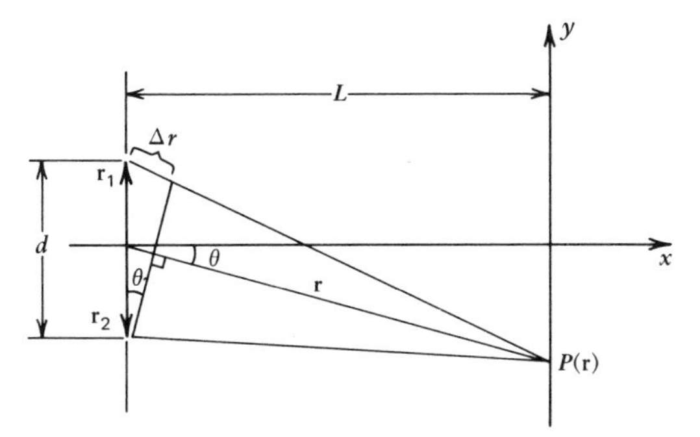

Figure 4.5. Geometry of wave paths in Young's experiment.

If the separation L between the screens is much greater than the separation between the pinholes d and if $P(\mathbf{r})$ remains in some reasonable region on the screen about the x axis, the two terms $|\mathbf{r} - \mathbf{r}_1|$ and $|\mathbf{r} - \mathbf{r}_2|$ become slowly varying and can be absorbed into the amplitude. In particular if we confine the analysis to the xy plane, we have

$$|\mathbf{r} - \mathbf{r}_1| = \left[L^2 + \left(y + \frac{d}{2}\right)^2\right]^{1/2} \simeq L\left[1 + \frac{1}{2}\frac{(y + d/2)^2}{L^2}\right] \tag{4.4a}$$

and

$$|\mathbf{r} - \mathbf{r}_2| = \left[L^2 + \left(y - \frac{d}{2}\right)^2\right]^{1/2} \simeq L\left[1 + \frac{1}{2}\frac{(y - d/2)^2}{L^2}\right]$$

making the path difference between the two waves

$$\Delta r = |\mathbf{r} - \mathbf{r}_1| - |\mathbf{r} - \mathbf{r}_2| = \frac{1}{2L}\left[\left(y + \frac{d}{2}\right)^2 - \left(y - \frac{d}{2}\right)^2\right] \tag{4.4b}$$

or

$$\Delta r = \frac{yd}{L}$$

The waves arriving at $P(x, y)$ can be expressed as

$$f(\mathbf{r}, t) = f_1 + f_2$$

$$= u_1 A_1 \cos\left(\frac{2\pi}{\lambda}|\mathbf{r} - \mathbf{r}_1| - \omega t\right) + u_2 A_2 \cos\left(\frac{2\pi}{\lambda}|\mathbf{r} - \mathbf{r}_2| - \omega t\right) \tag{4.5}$$

where the phases ϕ_1 and ϕ_2, which are fixed by the original source and the geometry to the pair of pinholes, are for convenience here set to zero. The only phase difference in the waves at $P(x, y)$ becomes

$$\Delta\phi = \frac{2\pi}{\lambda}(|\mathbf{r} - \mathbf{r}_1| - |\mathbf{r} - \mathbf{r}_2|) = \frac{2\pi}{\lambda}\Delta r = \frac{2\pi y d}{L\lambda} \tag{4.6}$$

Neglecting diffraction we determine the position of the maxima and minima of the intensity pattern on the screen by

$$\frac{2\pi y d}{L\lambda} = 2\pi n \quad \text{or} \quad y = \frac{n\lambda L}{d} \qquad \text{for maxima} \tag{4.7}$$

$$\frac{2\pi y d}{L\lambda} = (2n + 1)\frac{\pi}{2} \quad \text{or} \quad y = \frac{(2n + 1)\lambda L}{d} \qquad \text{for minima}$$

where $n = 0, 1, 2, \ldots$, thereby providing a central maximum located at $y = 0$ with alternating variations in the intensity along the y axis. Another way of achieving this result is to note from Fig. 4.5 that

$$\tan\theta = \frac{y}{L} \quad \text{and} \quad \sin\theta = \frac{\Delta r}{d} \tag{4.8}$$

In the approximation $y \ll L$ and $\Delta r \ll d$ we can express $\sin\theta \simeq \tan\theta \simeq \theta$, making again

$$\Delta r = \frac{yd}{L} \tag{4.9}$$

but this time also showing that the path difference must be small compared to the separation of the pinholes.

4.2 INTENSITY OF THE FRINGES

Representing the composite wave at the point $P(x, y)$ as

$$f(\mathbf{r}, t) = \mathrm{Re}\left[u_1 A_1 e^{i(k|\mathbf{r}-\mathbf{r}_1| - \omega t + \phi_1)} + u_2 A_2 e^{i(k|\mathbf{r}-\mathbf{r}_2| - \omega t + \phi_2)}\right] \tag{4.10}$$

we have included the $\mathbf{r}$ dependence of the spherical waves in the amplitudes and for generality expressed the initial phases at the pinholes as ϕ_1 and ϕ_2. With the exponentials rewritten as

$$f(\mathbf{r}, t) = \mathrm{Re}\left[\left(u_1 A_1 e^{i\delta_1} + u_2 A_2 e^{i\delta_2}\right)e^{-i\omega t}\right] \tag{4.11}$$

where

$$\delta_1 = k|\mathbf{r} - \mathbf{r}_1| + \phi_1 \quad \text{and} \quad \delta_2 = k|\mathbf{r} - \mathbf{r}_2| + \phi_2$$

The composite wave becomes

$$f(\mathbf{r}, t) = \mathrm{Re}(Ce^{i\delta}e^{-i\omega t}) = C\cos(\delta - \omega t) \tag{4.12a}$$

where

$$C^2 = (u_1A_1)^2 + (u_2A_2)^2 + 2(u_1A_1)(u_2A_2)\cos(\delta_1 - \delta_2) \tag{4.12b}$$

Since the intensity is related to the square of the wave function $I = |f(\mathbf{r}, t)|^2$ we have the instantaneous intensity

$$I(t) = \left[(u_1A_1)^2 + (u_2A_2)^2 + 2u_1u_2A_1A_2\cos(\delta_1 - \delta_2)\right]\cos^2(\delta - \omega t) \tag{4.13}$$

dictating the pattern on the screen.

We have already shown the average of $\cos^2\theta$ over any number of complete cycles to have the value one-half. The average value over any time τ for Eq. (4.13) is

$$\langle I\rangle_{\text{time}} = \frac{C^2}{\tau}\int_0^\tau \cos^2(\delta - \omega t)\,dt = \frac{C^2}{\tau}\int_0^\tau \frac{1 + \cos 2(\delta - \omega t)}{2}\,dt$$

$$= \frac{C^2}{2} - \frac{C^2}{4\omega\tau}\sin 2(\delta - \omega\tau) - \sin\delta \tag{4.14}$$

Optical frequencies are in the range of 10^{15} Hz making ω quite large. Taking τ to be the response time of a measuring instrument recording the intensity, the quantity $\omega\tau$ will be much greater than one, since the best of detectors are still slow compared to optical periods ($\tau > 10^{-12}$ s). This leaves us no hope of following an individual cycle let alone even a small number of cycles, making the time-averaged intensity of a sinusoidal wave

$$\langle I\rangle_{\text{time}} = \frac{C^2}{2} = I_0 \tag{4.15}$$

Therefore, the time average of Eq. (4.13) over any time τ, where $\omega\tau \gg 1$, leads to the time-independent expression

$$I_0 = I_1 + I_2 + 2\sqrt{I_1 I_2}\cos(\delta_1 - \delta_2) \tag{4.16}$$

The intensity on the screen is the sum of the intensities of each individual wave plus a third term referred to as the *interference term*. As the path difference Δr varies, $(\delta_1 - \delta_2)$ will change allowing $\cos(\delta_1 - \delta_2)$ to sweep through all values from plus to minus one. The interference term

$$2\sqrt{I_1 I_2}\cos(\delta_1 - \delta_2) \tag{4.17}$$

varies with the position $P(x, y)$ on the screen, giving extreme values for the intensity

$$I_{0\,\substack{\max\\\min}} = I_1 + I_2 \pm 2\sqrt{I_1 I_2} \tag{4.18}$$

If the two slits of Young's experiment provide identical diffraction effects, the coefficients u_1 and u_2 will be equal and can be considered to contribute a coefficient $u = u_1 = u_2$ to the combined wave. If we assume the amplitudes of the individual waves also to be equal, $A_1 = A_2$, then, from Eq. (4.12b)

$$I_0 = \frac{C^2}{2} = 2I_1[1 + \cos(\delta_1 - \delta_2)] \tag{4.19}$$

Finally, if the initial phases of the individual waves are equal, we have

$$\delta_1 - \delta_2 = \frac{2\pi y d}{L\lambda}.$$

making

$$I_0 = 2I_1\left(1 + \cos\frac{2\pi y d}{L\lambda}\right) \tag{4.20}$$

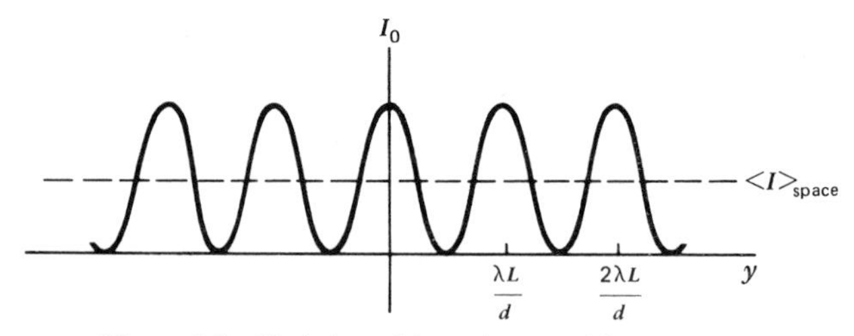

Figure 4.6. Variation of intensity over fringe pattern.

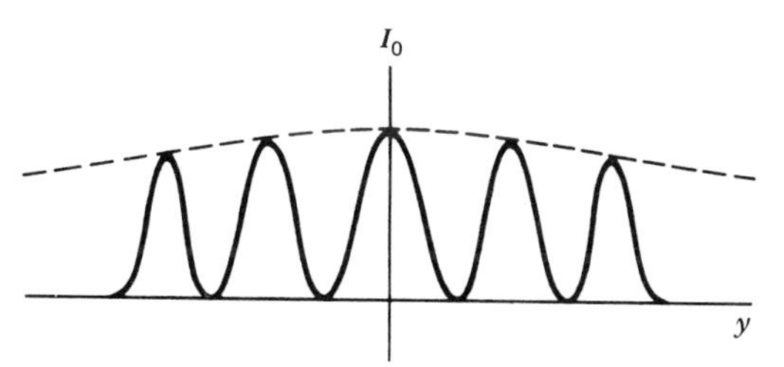

Figure 4.7. Modification of intensity due to diffraction.

With the assumption that $u = 1$ (point sources), a plot of the intensity (Fig. 4.6) shows the location of the maximum and zero intensities. It also shows that a space average of the intensity over any whole number of cycles of the fringe pattern is $2I_1$, as we would expect from the conservation of energy. The interference pattern redistributes the available energy providing a maximum of $4I_1$ at some locations and zero at others. If the diffraction effect is included, a slit of width $w \ll d$ will produce a slowly varying envelope for the interference pattern in the form[‡]

$$u^2 = \frac{\sin^2(\pi wy/L\lambda)}{(\pi wy/L\lambda)^2} \tag{4.21}$$

making a more realistic pattern of the intensity distribution appear, as in Fig. 4.7.

The analysis confirms the experimental results of the Young interference fringe pattern. In obtaining the prediction, however, we have assumed perfect sinusoidal waves throughout. It remains to investigate light from a real source and understand the role it plays in interference effects.

4.3 SOURCES OF CHAOS

A conventional light source consists of a large number of atomic oscillators, each radiating a limited wave pulse at random times. The sun or an incandescent light bulb represent a thermal source having a broad continuous spread in frequencies. A gas discharge tube containing a low vapor pressure of a substance activated by a high enough voltage across the gas is another random source of light. The latter is distinguished by a discontinuous spread in frequencies that tend to cluster about particular mean

[‡]See Ditchburn, R. W. *Light*, Interscience, New York, 1955, p. 158.

frequencies giving the appearance of a line spectrum. The atoms in any of these sources gain energy through collisions and absorption of light and then spontaneously radiate this energy at random times from random positions with random velocities.

Although the collisions can be somewhat diminished by reducing the temperature of the source and the emissions can be filtered to extremely narrow ranges of frequency, the light will always be characterized by the chaotic manner in which the independent atomic oscillators radiate. Until recently these thermal or Gaussian sources were the only type that existed. With the advent of the laser, however, an optical source became available in which the emissions of individual atoms could be coordinated in phase to produce light of a nonrandom, coherent nature. The laser is inherently a quantum source to be dealt with in later chapters.

Typically the emissions from a single atom occur in an average decay time of about 10^{-8} s. The radiation in this time interval τ_d may be represented by a wave packet of central angular frequency ω_0. Although it is impossible to know the precise nature of the emission, radiation from a single atom can be assumed to have a narrow spread in frequency and slowly varying amplitude. To a good approximation, we can represent the wave as

$$f(t) = \dot{A}e^{i(\omega_0 t + \phi)} \qquad 0 \leqslant t \leqslant \tau_d \tag{4.22}$$

where ϕ is a constant phase within the decay interval τ_d (Fig. 4.8).

An understanding of the line broadening or frequency spread for this single truncated wave can be gained by considering the Fourier transform for this function. In the previous chapter Fourier theory was formulated for spatial, time-independent wave components e^{ikx}. Here we are interested in the time-dependent part $e^{i\omega t}$. By associating ω with k and t with x the

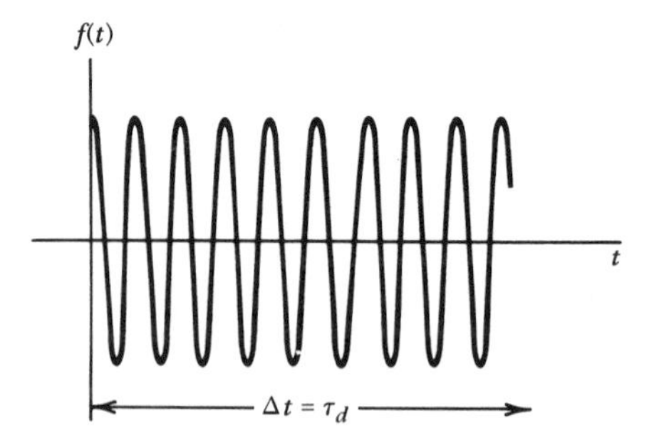

Figure 4.8. Truncated plane wave.

Fourier transforms become

$$G(\omega) = \frac{1}{\sqrt{2\pi}} \int_{-\infty}^{\infty} F(t) e^{-i\omega t}\, dt$$

and

$$F(t) = \frac{1}{\sqrt{2\pi}} \int_{-\infty}^{\infty} G(\omega) e^{i\omega t}\, d\omega \tag{4.23}$$

With these integrals applied to the truncated plane wave (4.22), where for convenience ϕ is set to zero and the time scale is adjusted so that

$$F(t) = A e^{i\omega_0 t} \qquad -\frac{\tau_d}{2} \leqslant t \leqslant \frac{\tau_d}{2} \tag{4.24}$$

then

$$G(\omega) = \frac{A}{\sqrt{2\pi}} \int_{-\tau_d/2}^{\tau_d/2} e^{i(\omega_0 - \omega)t}\, dt$$

$$= A\sqrt{\frac{2}{\pi}}\, \frac{\sin(\omega_0 - \omega)\tau_d/2}{\omega_0 - \omega}$$

showing the emission from a single atom to have a natural spread in frequency related to the limited time of radiation, and as was previously established for spatial transforms

$$\Delta\omega\, \Delta t \simeq 1 \tag{4.25}$$

The frequency spread is related to the characteristic time of the truncated sinusoidal wave which in this case is the decay time

$$\tau_d = \Delta t = \frac{1}{\Delta\omega} \tag{4.26}$$

This natural line broadening, therefore, is inherent in the very emission process itself and will be present even for laser light. Ultimately it is a quantum effect limited by the precision with which the energy of the radiation is known or by the time interval within which a single atom can radiate. Luckily, it is a small effect and can be omitted at this point without any loss in generality. Since it contributes an amplitude modulation it can always be added later.

At optical frequencies of the order of 10^{15} Hz an undisturbed atomic emission contains approximately 10^7 cycles. Ideally we wish this atom to be at rest or have a small velocity. Generally this is not the case, since the atom is part of a thermal system of particles having random velocities. Thus another effect leading to line broadening is the Doppler effect in which the frequency of emission could be higher or lower pending the direction of motion of the emitting atom. The shift in frequency is determined by the velocity. Since this is a temperature-dependent effect, it can at least be kept small.

Finally collisions themselves contribute to the frequency spread by possibly chopping the wave from a single atom into smaller segments. If the average time between collisions in a source is less than the decay time for a single atom, the frequency spread $\Delta\omega$ become larger than the natural line broadening. A typical source might have a mean collision time of the order of 10^{-11} s where, on the average, a thousand collisions might occur during the radiation from a single atom. Each segment of radiation between collisions would contain about 10,000 cycles of the wave, still long enough to be represented as a sinusoidal oscillation over the characteristic time between collisions, τ_c. As with the Doppler effect, the collision broadening can be reduced by controlling the temperature of the source. The frequency spread for all three effects—Doppler, collision, and natural line broadening—can be minimized but will be present in any kind of radiation including laser light. Each effect will contribute at most an amplitude modulation to the original assumed sinusoidal radiation.

It is the collision effect, however, which provides an abrupt change that sets thermal light apart from laser light. A sudden random phase shift occurs with each collision (Fig. 4.9), so that even the radiation from a single atom has a constant phase at best only for the average time τ_c, the characteristic time between collisions. For any time greater than τ_c, the emitted wave has no memory of its former phase. Thus, the phase is

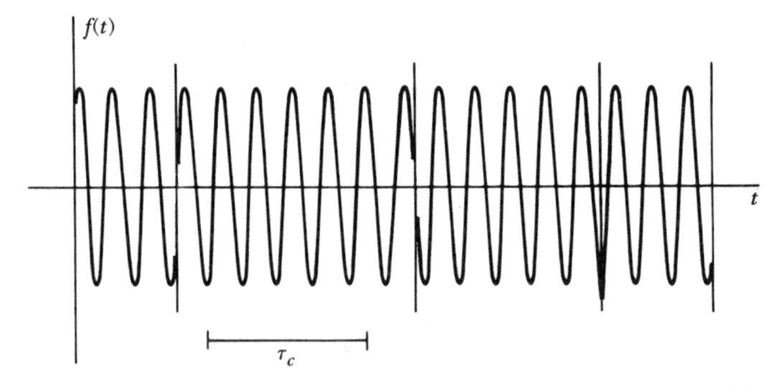

Figure 4.9. Wave segments of random phase. Vertical lines denote collisions.

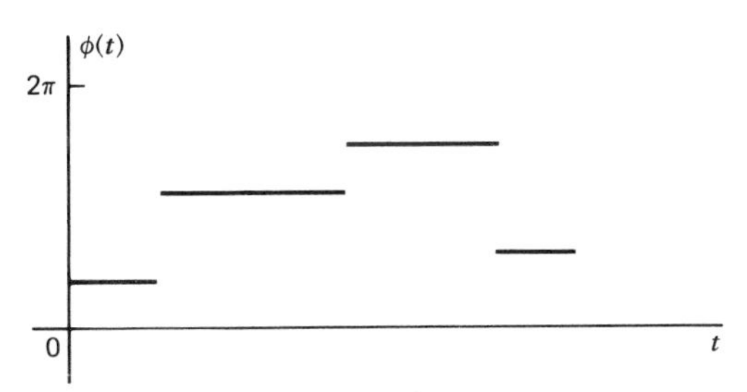

Figure 4.10. Randomly changing phase at each collision.

time-dependent, and over times much longer than the characteristic time τ_c, the phases $\phi(t)$ form a random set changing chaotically as pictured in the step function of Fig. 4.10. This single feature of unpredictable and uncontrollable phase shifts characterizes Gaussian light.

4.4 THE GAUSSIAN CHARACTER

A typical source, even small enough to be considered approximately a point source, contains an enormous number of atoms emitting at many different central frequencies ω_i. The subset N of just those atoms radiating at one particular frequency ω_0 is still large. At any instant, the composite wave of central frequency ω_0 is the sum of the contributions from each individual oscillator of that frequency. But all the oscillators are independent of one another and begin their oscillations at random times, giving each wave a different phase $\phi_n(t)$. The composite wave then is

$$F(t) = \sum_{n=1}^{N} f_n(t) = \sum_{n=1}^{N} A_n e^{i\omega_0 t} e^{i\phi_n(t)} \tag{4.27}$$

If each oscillator produces a wave of equal amplitude $A_n = A$, the total wave of single frequency becomes

$$F(t) = Ae^{i\omega_0 t} \sum_{n=1}^{N} e^{i\phi_n(t)} \tag{4.28}$$

At a given instant the sum of exponentials of random phases and unit magnitude is best viewed as a vector sum yielding some resultant vector of magnitude $a(t)$ at a phase angle $\delta(t)$, as in Fig. 4.11, namely,

$$a(t)e^{i\delta(t)} = \sum_{n=1}^{N} e^{i\phi_n(t)} \tag{4.29}$$

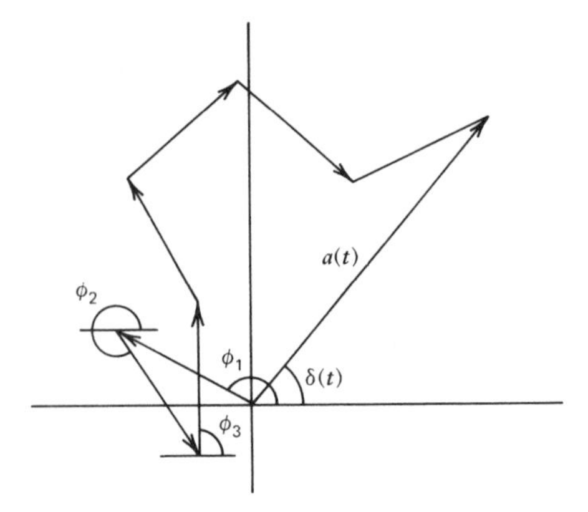

Figure 4.11. Sum of exponentials of random phase.

The amplitude $a(t)$ may vary between zero (which is highly probable since it can occur in many ways) to a maximum of N (which is highly improbable since it can occur in only one way, that is, all the random $\phi_n(t)$ equal).

This distribution is usually referred to as the *random walk problem*, where steps of unit length are taken one after the other in random directions. After N steps we can calculate the average distance wandered from the origin (an ensemble average of many repeated trials). The problem is similar to a set of unit radial vectors $\hat{\mathbf{r}}_n$ each set head to tail, capable of rotating to any direction carrying all succeeding vectors in the rotation. The first step must be unit distance $d_1 = 1$ from the origin. Since the second step is as likely to double the distance as to bring the random walker back to the origin, we will assume an average somewhere in between, namely, at right angles to the first step making the average distance progressed, $d_2 = \sqrt{2}$, (Fig. 4.12). If we allow the next step on the average to be a right angle to the line d_2 for the same reasoning, d_3 becomes $\sqrt{3}$. After N steps we have progressed on the average a distance $\sqrt{N}$ from the origin. Thus, the average value taken over many trials, or in the case of $a(t)$ over a long time, is $\langle a(t) \rangle = \sqrt{N}$. The direction, however, remains completely unpredictable.

As we have noted, the amplitude $a(t)$ may fluctuate between zero and N, each value having a probability of occurrence $p[a(t)]$. Since $a(t)$ has the same probability of falling within any differential area (Fig. 4.13), $2\pi a(t)\,da(t)$, we can define an incremental probability of finding $a(t)$ at a given instant as

$$dp \sim \frac{2\pi a(t)\,da(t)}{\pi \langle a(t) \rangle^2} \tag{4.30}$$

that is, the odds of finding $a(t)$ within some differential area relative to the

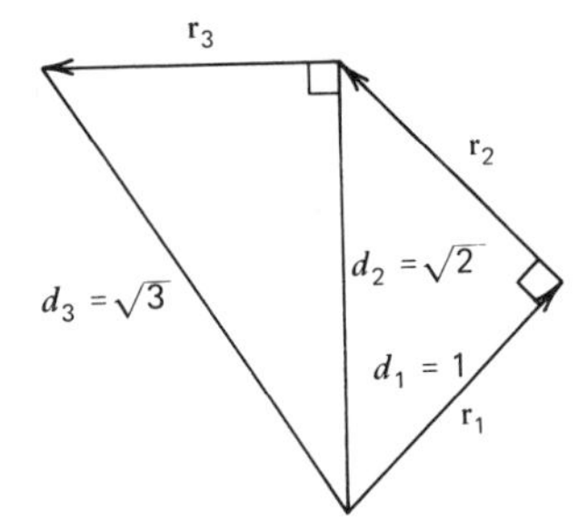

Figure 4.12. Construction for calculating average distance from origin.

average area of the average value of $a(t)$. But we have seen earlier that the probability $p[a(t)]$ is largest for $a(t) = 0$, and grows smaller as $a(t)$ increases. For a constant area, therefore, the change in probability should be proportional to the probability and opposite in direction.

$$dp \sim -p \tag{4.31}$$

Combining these assumptions we have

$$dp = -p\frac{2a(t)\,da(t)}{\langle a(t)\rangle^2} \tag{4.32}$$

or

$$\int_{p(0)}^{p[a(t)]}\frac{dp}{p} = -2\int_0^{a(t)}\frac{a(t)\,da(t)}{\langle a(t)\rangle^2}$$

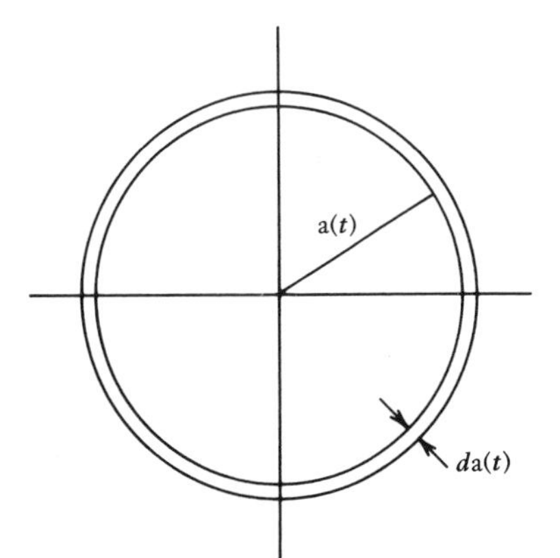

Figure 4.13. Differential area in random walk plane.

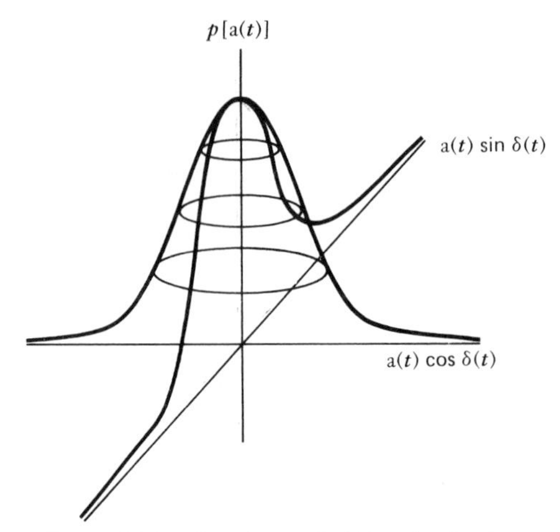

Figure 4.14. Gaussian probability distribution.

yielding

$$p[a(t)] = p(0)e^{-a(t)^2/\langle a(t)\rangle^2}$$

which is a Gaussian function depicting the probability distribution shown in Fig. 4.14. As anticipated, it is a symmetrical function independent of the phase $\delta(t)$ and has a most probable value for $a(t)$ equal to zero, where $p[a(t)]$ is a maximum.

The function can be normalized by making the total volume under the surface equal to one, which is the same as making the probability of finding $a(t)$ in the entire plane unity:

$$\int_0^{\infty} p[a(t)]2\pi a(t)\, da(t) = 1 \tag{4.33}$$

or

$$1 = 2\pi p(0)\int_0^{\infty} a(t)e^{-a(t)^2/\langle a(t)\rangle^2}\, da(t) = N\pi p(0)$$

where $N = \langle a(t)\rangle^2$, making

$$p(0) = \frac{1}{N\pi}$$

It is this probability function

$$p[a(t)] = \frac{1}{N\pi}e^{-a(t)^2/N} \tag{4.34}$$

that typifies all chaotic sources and from which the name Gaussian light derives. Thus while the random amplitude modulation $a(t)$ has an average value of $\sqrt{N}$ and a high probability of small values for $a(t)$, the amplitude can fluctuate wildly as the phases $\phi_n(t)$ change.

4.5 INTENSITY OF RANDOM LIGHT

The instantaneous wave of a single mode or frequency for a Gaussian source is

$$F(t) = Aa(t)e^{i\omega_0 t}e^{i\delta(t)} \tag{4.35}$$

for which $a(t)$ and $\delta(t)$ vary slowly within the time of a single cycle of the optical wave, approximately 10^{-15} s. A time average, therefore, of $|F(t)|^2$ over one cycle yields the average cycle intensity.

$$I(t) = \langle |F(t)|^2 \rangle_{1\text{ cycle}} = A^2 a(t)^2 \tag{4.36}$$

For a Gaussian distribution of the time-dependent amplitudes, however, this intensity is still time dependent, and can fluctuate broadly for times greater than the time of a few cycles. The phases $\phi_n(t)$ begin to change appreciably, and by the time τ has passed (decay time or mean collision time, whichever is less) practically all the individual phases have changed, leaving the values of both $a(t)$ and $\delta(t)$ unrelated to any former value of the wave. A plot of the cycle-averaged intensity might appear as in Fig. 4.15.

Generally, a measurement of the intensity occurs over times much longer than the characteristic time of phase changes and therefore averages the fluctuations. We have already seen from the random walk that $a(t)$ has an average value of $\sqrt{N}$ over many different trials. This is an ensemble average where the same experiment is repeated many times and an average is taken of the results. In the case of a steady light source where the long-time average of the intensity is constant (which is the same as saying that a sampling of the intensity in any long time interval is the same even for different times) the source is called a *stationary source*. A time average can then be used to replace the ensemble average. Performing this long-time average, we have

$$\langle |a(t)|^2 \rangle_{\text{time}} = \left\langle \sum_{n=1}^{N} e^{i\phi_n} \sum_{m=1}^{N} e^{-i\phi_m} \right\rangle \tag{4.37}$$

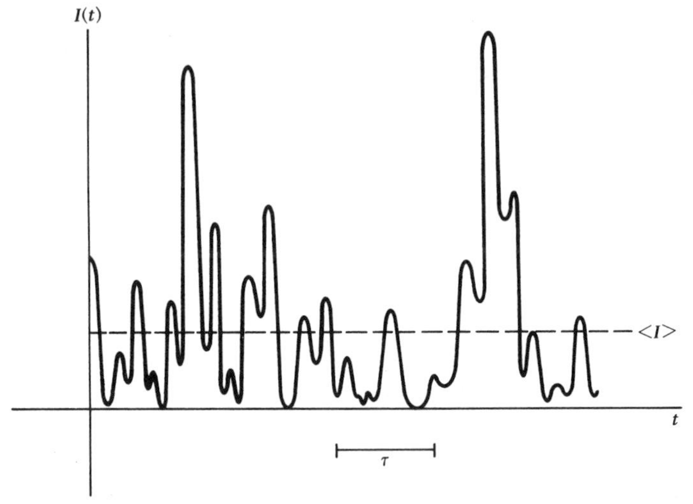

Figure 4.15. Fluctuation of cycle-averaged intensity.

Since there are N terms for which $m = n$ and $N(N - 1)$ terms for which $n \neq m$ we can write

$$\langle |a(t)|^2 \rangle = N + \left\langle \sum_{\substack{n=1 \\ n \neq m}}^{N} \sum_{m=1}^{N} e^{i(\phi_n - \phi_m)} \right\rangle \tag{4.38}$$

which can also be expressed, since n and m are symmetrical, as

$$\langle |a(t)|^2 \rangle = N + 2\left\langle \sum_{n>m=1}^{N} \cos(\phi_n - \phi_m) \right\rangle \tag{4.39}$$

But the phases ϕ_n vary in a purely random manner over a long-time average, allowing the cosine function to have as equal probability of being positive as negative. The last terms, therefore, vanish leaving the long-time average of the amplitude squared equal to N. In Fig. 4.15, this is seen as the long-time-averaged intensity

$$\langle I \rangle = NA^2 \qquad \text{for} \quad t \geqslant \tau \tag{4.40}$$

which would be the resultant measurement of a typical detector having a response time at best in the range of the characteristic time of the Gaussian light. Thus the fluctuations, with a high probability of being zero and a low probability of being as high as N^2A^2, would ordinarily escape detection.

If we take the characteristic time during which most of the phases ϕ_n change as the coherence time τ_c, two measurements of intensity made within the coherence time would have some relationship to one another, while two measurements made at different times greater than the coherence time carry no memory of the phases of one another. Using the concept of coherence time as the least time for which on the average all the phases ϕ_n change, we have an associated coherence length

$$\lambda_c = c\tau_c \tag{4.41}$$

where c is the speed of light, within which the wave still has at least some history of its former phase.

4.6 CORRELATION AND COHERENCE

If we consider light from a Gaussian point source traveling along two different paths and then recombining at a point, the combined wave (Fig. 4.16) is

$$F(\mathbf{r}, t) = F(\mathbf{r}_1, t_1) + F(\mathbf{r}_2, t_2) \tag{4.42}$$

where $t_1 = t - s_1/c$ and $t_2 = t - s_2/c$. This is similar to Young's experiment or any interference experiment, where light from the same point source is split in two parts and recombined after one beam has traveled a different distance (or what is equivalent, a different time) than the other beam. In particular $|\mathbf{r}_1| = |\mathbf{r}_2|$ in a symmetrical arrangement of Young's experiment, so that as before

$$|\mathbf{r} - \mathbf{r}_1| = s_1 \quad \text{and} \quad |\mathbf{r} - \mathbf{r}_2| = s_2 \tag{4.43}$$

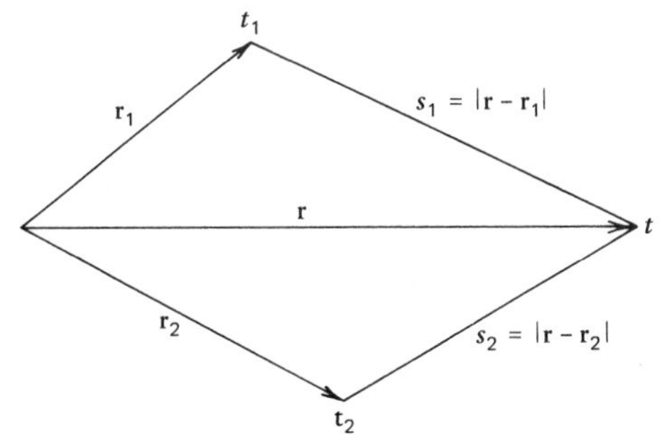

Figure 4.16. Time–space geometry of two paths for light waves.

The shift in the space–time coordinates of the beams at the point $(\mathbf{r}, t)$ becomes

$$\tau = t_2 - t_1 = \frac{s_1 - s_2}{c} \tag{4.44}$$

making it equivalent to speak of either a time delay τ for the two beams or a difference in path length $s_1 - s_2$ for the two beams.

Still another way of looking at the sum (4.42) is to consider that $F(\mathbf{r}, t)$ is a sampling of the wave front at two different space–time points $(\mathbf{r}_1, t_1)$ and $(\mathbf{r}_2, t_2)$. The intensity of the combined wave averaged over one cycle is

$$I(\mathbf{r}, t) = F^*(\mathbf{r}, t)F(\mathbf{r}, t) = |F(\mathbf{r}, t)|^2 \tag{4.45}$$

where again the time dependence is retained for the intensity as in the case for chaotic light having random phases, and therefore a time-dependent amplitude. With the intensity at $\mathbf{r}$ and t expressed in terms of the individual waves

$$\begin{aligned} I(\mathbf{r}, t) &= |F(\mathbf{r}_1, t_1) + F(\mathbf{r}_2, t_2)|^2 \\ &= |F(\mathbf{r}_1, t_1)|^2 + |F(\mathbf{r}_2, t_2)|^2 + 2\,\mathrm{Re}\,F^*(\mathbf{r}_1, t_1)F(\mathbf{r}_2, t_2) \end{aligned} \tag{4.46}$$

where the third term takes its form from the definition (3.21), namely,

$$\mathrm{Re}\,F^*(\mathbf{r}_1, t_1)F(\mathbf{r}_2, t_2) = \frac{F^*(\mathbf{r}_1, t_1)F(\mathbf{r}_2, t_2) + [F^*(\mathbf{r}_1, t_1)F(\mathbf{r}_2, t_2)]^*}{2} \tag{4.47}$$

A long-time average of the intensity, Eq. (4.46), becomes

$$\langle I \rangle = \langle I_1 \rangle + \langle I_2 \rangle + 2\,\mathrm{Re}\langle F^*(\mathbf{r}_1, t_1)F(\mathbf{r}_2, t_2)\rangle \tag{4.48}$$

The time average is performed over a time greater than τ_c to conform with the realities of experiments. Whether Young's experiment or any other first order (two wave fields) experiment is performed where the fringes are recorded by the eye, a photographic plate or photodetector, the pattern observed or measured is this long-time average. Expressed more explicitly

$$\langle F^*(\mathbf{r}_1, t_1)F(\mathbf{r}_2, t_2)\rangle = \lim_{T\to\infty} \frac{1}{T}\int_0^T F^*(\mathbf{r}_1, t_1)F(\mathbf{r}_2, t_1 + \tau)\, dt_1 \tag{4.49}$$

where $t_2 = t_1 + \tau$. Finally, this average is referred to as the first-order correlation function, and the extent to which it is not zero indicates that at least some interference can take place and that the two wave fields show some coherence relative to one another. Normalizing the correlation function we have

$$g^{(1)}(\mathbf{r}_1, t_1; \mathbf{r}_2, t_2) = \frac{\langle F^*(\mathbf{r}_1, t_1) F(\mathbf{r}_2, t_2) \rangle}{\langle |F(\mathbf{r}_1, t_1)|^2 \rangle^{1/2} \langle |F(\mathbf{r}_2, t_2)|^2 \rangle^{1/2}} \tag{4.50}$$

In this way $|g^{(1)}|$ will range in value from zero to one, corresponding to first-order incoherence at zero through partial coherence to full first-order coherence at the value one. The function for stationary light sources is independent of time, depending only upon τ, the difference in time of arrival of the two waves. It is this function we wish to evaluate for Gaussian light from the pinholes in Young's experiment in the form

$$F(\mathbf{r}_1, t_1) = \sum_{n=1}^{N} A e^{i[\omega t_1 + \phi_n(t_1)]} = A e^{i(\omega t - k|\mathbf{r} - \mathbf{r}_1|)} \sum_{n=1}^{N} e^{i\phi_n(t_1)} \tag{4.51}$$

and

$$F(\mathbf{r}_2, t_2) = \sum_{m=1}^{N} A e^{i[\omega t_2 + \phi_m(t_2)]} = A e^{i(\omega t - k|\mathbf{r} - \mathbf{r}_2|)} \sum_{m=1}^{N} e^{i\phi_m(t_2)}$$

A calculation of the time averages reduces to performing the integral

$$\frac{1}{T}\int_0^T F^*(\mathbf{r}_1, t_1) F(\mathbf{r}_2, t_2)\, dt_1 = \frac{1}{T} \sum_{m,n=1}^{N} A^2 e^{ik(s_1 - s_2)} \int_0^T e^{i[\phi_m(t_1+\tau) - \phi_n(t_1)]}\, dt_1 \tag{4.52}$$

For $n \neq m$, the phases form a random set of differences which average to zero. Only when $n = m$, do we deal with the same phase but at a different time.

A single phase $\phi_n(t)$ changes on the average within a time τ_c. When we consider the same function $\phi_n(t_1)$ and $\phi_n(t_1 + \tau)$, shifted on the time scale, as in Fig. 4.17, the difference between the time displaced phase (Fig. 4.18) shows that where there is overlap, the difference is zero; otherwise, the difference with the next random change in phase is again random. Thus for

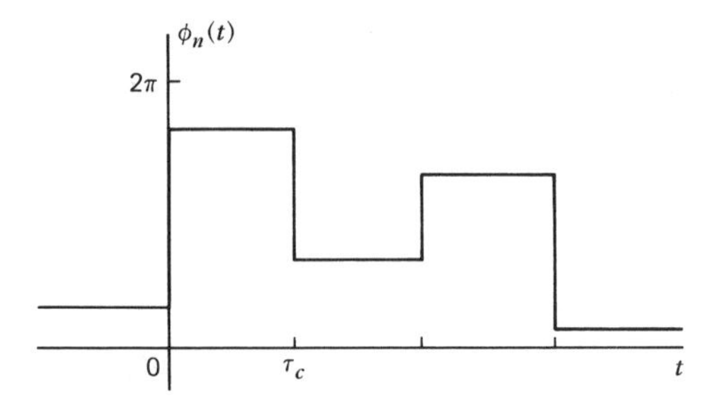

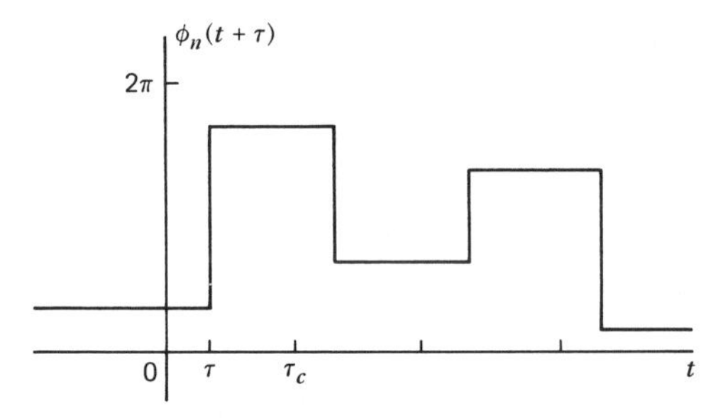

Figure 4.17. Plot of random phase $\phi_n(t)$ and $\phi_n(t+\tau)$.

$m = n$, the integral (4.52), for one time interval τ_c, becomes

$$\frac{1}{\tau_c}\left[\int_0^{\tau} e^{i\delta_n}\,dt_1 + \int_{\tau}^{\tau_c} dt_1\right] = e^{i\delta_n}\frac{\tau}{\tau_c} + \frac{\tau_c - \tau}{\tau_c} \tag{4.53}$$

where $\delta_n = \phi_n(t_1) - \phi_n(t_1 + \tau)$. Successive integration over many time intervals yields a similar result giving a collection of random terms $e^{i\delta_n}$ (which average to zero) and the average term $(\tau_c - \tau)/\tau_c$. Since there are N terms for which $m = n$, the time average, Eqs. (4.49) and (4.52), is

$$\langle F^*(\mathbf{r}_1, t_1)F(\mathbf{r}_2, t_2)\rangle = NA^2\frac{\tau_c - \tau}{\tau_c}e^{ik(s_1 - s_2)} \tag{4.54}$$

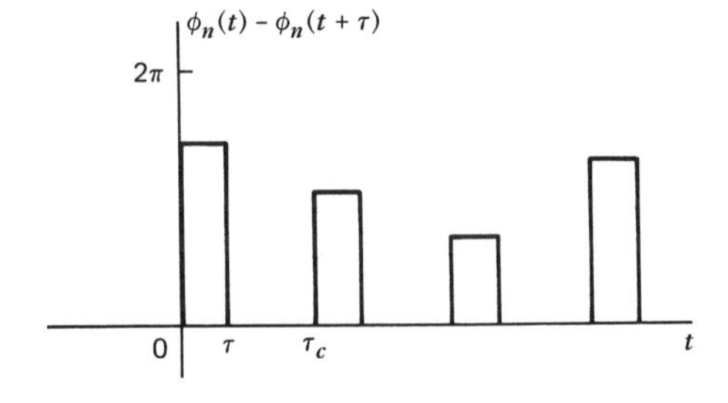

Figure 4.18. A plot of the difference in phase.

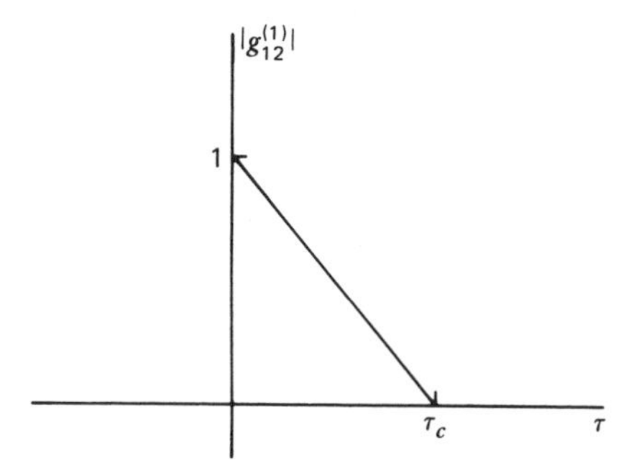

Figure 4.19. The normalized first-order correlation function for Gaussian light.

making the interference term from Eq. (4.48) become

$$2\,\mathrm{Re}\langle F^*(\mathbf{r}_1, t_1)F(\mathbf{r}_2, t_2)\rangle = 2NA^2\left(1 - \frac{\tau}{\tau_c}\right)\cos k\,\Delta s \qquad (4.55)$$

If we use

$$\langle F^*(\mathbf{r}_1, t_1)F(\mathbf{r}_1, t_1)\rangle^{1/2} = \sqrt{N}\,A \qquad (4.56)$$

and

$$\langle F^*(\mathbf{r}_2, t_2)F(\mathbf{r}_2, t_2)\rangle^{1/2} = \sqrt{N}\,A$$

the normalized correlation function for this particularly simplified model of Gaussian light becomes (Fig. 4.19)

$$|g_{12}^{(1)}| = 1 - \frac{\tau}{\tau_c} \qquad \text{for } \tau < \tau_c \qquad (4.57)$$

If the time displacement τ is greater than τ_c, then not even the same phase ϕ_n (Fig. 4.18) overlaps at all with itself, making the time-averaged integral zero at all times; thus

$$|g_{12}^{(1)}| = 0 \qquad \tau \geqslant \tau_c \qquad (4.58)$$

We see then that chaotic light of a single frequency can be partially first-order coherent for $\tau < \tau_c$ and even fully first-order coherent for $\tau = 0$. The latter would be the case for the central fringe in Young's experiment. For the other fringes

$$\tau = \frac{\Delta s}{c} = \frac{n\lambda}{c} = \frac{n}{f} \qquad n = 1, 2, \ldots \qquad (4.59)$$

and since for an average thermal source, τ_c might be in the range of 10^{-12} s, we can expect up to a thousand fringes for light of frequency 10^{15} Hz. Of course, we have considered thus far only monochromatic light. Introducing line broadening of the three varieties mentioned earlier will reduce the number of fringes. A larger effect reducing the number is the fact that even small sources have some spatial extention making τ_c for the chaotic beam even smaller.

The stable plane wave assumed in Section 4.1 for Young's experiment yields $|g_{12}^{(1)}| = 1$, showing the full coherence representative of perfect laser light. In actuality a reasonably stable laser might exhibit a coherence time of 10^{-3} s which nonetheless corresponds to a correlation function that remains very close to unity.

Most classical interference experiments make use of the first-order coherence properties of Gaussian light. As long as path length differences or time shifts are less than the coherence length and time, reasonable interference results. Since most detectors respond in a time greater than the coherence time it is generally impossible to see the fluctuations of the chaotic beam. In a new breed of experiment performed by Hanbury Brown and Twiss,‡ an interference of *intensities* allows measurement of the product of intensity fluctuations. Although the experiment is intrinsically a quantum phenomenon since it depends on the counting and correlation of photons by photodetectors, it will be introduced here and discussed classically to provide an example of higher order correlation.

The intensity at two different space–time points is measured by photodetectors, as shown in Fig. 4.20. It must be assumed that the output of each detector is proportional to the intensity. One output is delayed by a time $t_2 - t_1 = \tau$, and two intensities are multiplied together as an interference effect and then averaged. With the high frequency of the original light removed (averaged) at the photodetector, the lower fluctuation frequency is used in the interference. In essence, the experiment makes a measurement of

$$\langle I(\mathbf{r}_1, t_1) I(\mathbf{r}_2, t_2) \rangle = \langle I \rangle^2 + \langle \Delta I(\mathbf{r}_1, t_1) \Delta I(\mathbf{r}_2, t_2) \rangle \tag{4.60}$$

which can be shown using

$$\langle I(\mathbf{r}_1, t_1) \rangle = \langle I(\mathbf{r}_2, t_2) \rangle = \langle I \rangle \tag{4.61}$$

and defining the fluctuation as

$$\Delta I(\mathbf{r}, t) = I(\mathbf{r}, t) - \langle I \rangle \tag{4.62}$$

‡Hanbury Brown, R. and R. Q. Twiss. *Nature* **177**, 27 (1956); Hanbury Brown, R. and R. Q. Twiss. *Proc. R. Soc. Lond. Ser. A* **243**, 291 (1957).

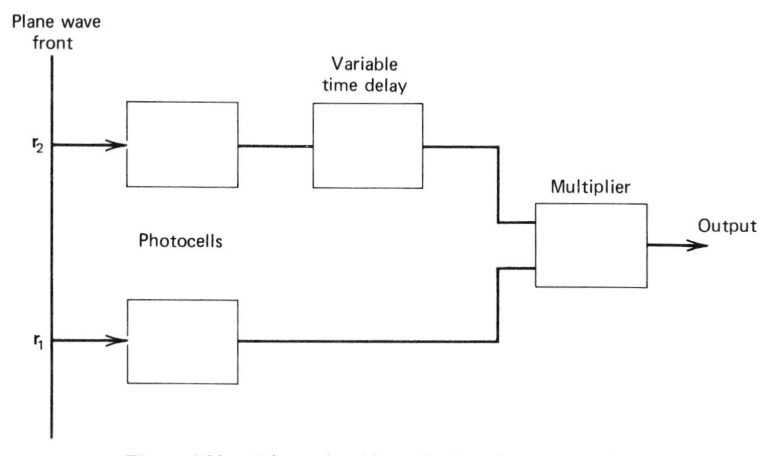

Figure 4.20. Schematic of intensity interference experiment.

where each intensity is the cycle-averaged intensity. Equation (4.60) can also be expressed as

$$\begin{aligned}\langle I(\mathbf{r}_1, t_1)I(\mathbf{r}_2, t_2)\rangle &= \langle |F(\mathbf{r}_1, t_1)|^2 |F(\mathbf{r}_2, t_2)|^2\rangle \\ &= \langle F^*(\mathbf{r}_1, t_1)F^*(\mathbf{r}_2, t_2)F(\mathbf{r}_2, t_2)F(\mathbf{r}_1, t_1)\rangle \quad (4.63)\end{aligned}$$

We have already shown for Gaussian light that the cycle-averaged intensity is

$$I(\mathbf{r}, t) = F^*(\mathbf{r}, t)F(\mathbf{r}, t) = A^2\left[N + \sum_{\substack{m,n=1\\m\neq n}}^{N} e^{i(\phi_m - \phi_n)}\right] \quad (4.64)$$

where normally the second term averages to zero for random phases. But here we are correlating one intensity with another, such that the product of the two intensities averaged over a long time becomes

$$\langle I(\mathbf{r}_1, t_1)I(\mathbf{r}_2, t_2)\rangle = A^4\left[N^2 + \left\langle \sum_{\substack{m,n=1\\m\neq n}}^{N} e^{i[\phi_m(t_2) - \phi_n(t_2)]} \sum_{\substack{i,j=1\\i\neq j}}^{N} e^{i[\phi_i(t_1) - \phi_j(t_1)]}\right\rangle\right]$$

(4.65)

where the cross terms average to zero:

$$\left\langle \sum_{\substack{m,n=1 \\ m \neq n}}^{N} e^{i(\phi_m - \phi_n)} \right\rangle = 0 \tag{4.66}$$

However, when $m = j$ and $n = i$, the product of the sums in Eq. (4.65) becomes

$$A^4 \left\langle \sum_{\substack{m=j=1 \\ m \neq n}}^{N} e^{i[\phi_m(t_1+\tau) - \phi_n(t_1)]} \sum_{n=i=1}^{N} e^{-i\phi_n(t_1+\tau) - \phi_n(t_1)} \right\rangle \tag{4.67}$$

Since $m \neq n$, the differences in phases $\Delta\phi_m$ and $\Delta\phi_n$ are each individually random and statistically independent of one another. All other combinations of terms, namely, $m \neq j$ and $n \neq i$ are random and average to zero. Therefore, the time average for Eq. (4.67) can be taken for each sum independently, and becomes

$$\langle F^*(\mathbf{r}_1, t_1) F(\mathbf{r}_2, t_2) \rangle^2 \tag{4.68}$$

making Eq. (4.65) take the final form

$$\langle I(\mathbf{r}_1, t_1) I(\mathbf{r}_2, t_2) \rangle = \langle I \rangle^2 + \langle F^*(\mathbf{r}_1, t_1) F(\mathbf{r}_2, t_2) \rangle^2 \tag{4.69}$$

Normalizing the averages we can define the second-order correlation function as

$$g_{12}^{(2)} = \frac{\langle I(\mathbf{r}_1, t_1) I(\mathbf{r}_2, t_2) \rangle}{\langle I(\mathbf{r}_1, t_1) \rangle \langle I(\mathbf{r}_2, t_2) \rangle} \tag{4.70}$$

with which Eq. (4.69) can be written as

$$g_{12}^{(2)} = 1 + g_{12}^{(1)^2} \tag{4.71}$$

Thus, a measure of the first-order correlation function for chaotic light yields the second-order value which has a maximum absolute value of 2. In the case of the plane wave of stable phase, namely, perfect laser light $|g_{12}^{(2)}| = 1$, the condition for perfect second-order coherence. This shows that chaotic light, which can be first-order coherent, cannot be second-order coherent, providing a deeper understanding of coherence and rendering dramatically the limitation of coherence for Gaussian light (Fig. 4.21).

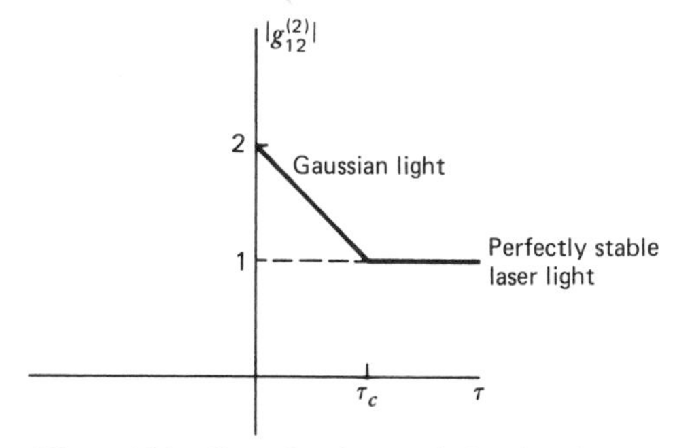

Figure 4.21. Second-order correlation function.

In principle, the second-order correlation function can be defined, most generally, for a sampling of wave fields from any four space–time points by

$$g^{(2)}(\mathbf{r}_1 t_1, \mathbf{r}_2 t_2; \mathbf{r}_3 t_3, \mathbf{r}_4 t_4)$$

$$= \frac{\langle F^*(\mathbf{r}_1, t_1) F^*(\mathbf{r}_2, t_2) F(\mathbf{r}_3, t_3) F(\mathbf{r}_4, t_4) \rangle}{(\langle |F(\mathbf{r}_1, t_1)|^2 \rangle \langle |F(\mathbf{r}_2, t_2)|^2 \rangle \langle |F(\mathbf{r}_3, t_3)|^2 \rangle \langle |F(\mathbf{r}_4, t_4)|^2 \rangle)^{1/2}} \tag{4.72}$$

The order of the wave functions is written to correspond to the formal quantum definition, where the complex functions become noncommuting operators whose order is fixed. A discussion of this and higher order correlation will be left to the quantum treatment since second-order correlation is best discussed in terms of photon counting and photon correlation relating statistical coincidences of photodetection.

Both atomic emission and energy measurement of waves can only be understood in quantum terms and photon properties. Thus, we are reaching the limitations of application of the wave model for light and must begin to investigate the mechanisms by which photon properties appear. A model of light based on these photon characteristics, then, must be applied to the experiments which we have discussed to predict results in agreement with current measurements. Ironically, we shall begin our understanding of light quanta with the very theory which almost irrefutably demonstrated the precise wave nature for light—the electromagnetic theory.

REFERENCES

Born, M. and E. Wolf. *Principles of Optics*, Pergamon, New York, 1970.

Ditchburn, R. W. *Light*, Interscience, New York, 1955.

Feynman, R. P., R. B. Leighton, and L. M. Sands. *The Feynman Lectures on Physics*, Vol. 1, Addison-Wesley, Reading, Mass., 1963.

Fowles, G. R. *Introduction to Modern Optics*, Rinehart and Winston, New York, 1968.

Jenkins, R. A. and H. E. White. *Fundamentals of Optics*, McGraw-Hill, New York, 1950.

Loudon, R. *The Quantum Theory of Light*, Oxford Clarendon Press, London, 1973.

Meyer-Arendt, J. R. *Introduction to Classical and Modern Optics*, Prentice-Hall, Englewood Cliffs, N.J., 1972.

Smith, F. G. and J. H. Thomson. *Optics*, Wiley, New York, 1971.

PROBLEMS

4.1. Show that $\mathrm{Re}(\mathbf{A}e^{-i\omega t}) \cdot \mathrm{Re}(\mathbf{B}e^{-i\omega t}) \neq \mathrm{Re}(\mathbf{A}e^{-i\omega t} \cdot \mathbf{B}e^{-i\omega t})$ where **A** and **B** are also complex quantities.

4.2. Prove that $\mathrm{Re}(\mathbf{A} \cdot \mathbf{B}^*) = \mathrm{Re}(\mathbf{A}^* \cdot \mathbf{B})$.

4.3. In Young's experiment, find the intensity at the center of the screen if a third slit were present midway between the two original ones.

4.4. Defining the term *fringe visibility* $V = (I_{\max} - I_{\min})/(I_{\max} + I_{\min})$, show that $V = 1$ in the case of complete coherence in Young's experiment.

4.5. For a Gaussian pulse $F(t) = Ae^{-\alpha t^2}e^{-i\omega_0 t}$, show that $G(\omega)$ is also Gaussian-shaped centered on the frequency ω_0.

4.6. Using the definition of the cycle-averaged intensity for a Gaussian source show that the probability function is

$$p[I(t)] = \frac{1}{\langle I \rangle} e^{-I(t)/\langle I \rangle}$$

4.7. Show that $\langle I(t)^n \rangle = n!\langle I \rangle$. This can be found by evaluating

$$\langle I(t)^n \rangle = \int_0^\infty p[I(t)]\, I(t)^n \, dI(t)$$

4.8. Using the approximation $L \gg d$, show Eqs. (4.4a) to be true.

4.9. From Fig. 4.12, show that, if a person takes random steps each of length L from some original point, after N steps that person would have progressed on the average a distance $\sqrt{NL}$ from the origin.

4.10. Carry out the integration indicated in Eq. (4.53).

4.11. Using the results of this integration, show that the interference term in Eq. (4.46) becomes $2NA^2(1 - \tau/\tau_c)\cos k\,\Delta s$.

4.12. With the result from Problem 4.11, derive the correlation function $|g_{12}^{(1)}|$ indicated in Eq. (4.57) for a time delay $\tau = \frac{1}{2}\tau_c$.

4.13. Show that Eq. (4.70) can be written as $g_{12}^{(2)} = 1 + g_{12}^{(1)^2}$.

5

Electromagnetic Waves and the Mechanics of Quanta

Electromagnetic theory provides the wave model of light with undulating electric and magnetic fields. Solutions of Clerk Maxwell's equations corroborate the known properties of wave propagation and predict the energy flow, interference effects, and physical interaction for optical waves. The phenomenon of light after Maxwell is synonymous with the behavior of electromagnetic fields.

It is ironic that in the very demonstrations conducted by Heinrich Hertz on the generation and transmission of electromagnetic waves, an observation emerged that would eventually challenge the wave theory and lead to a photon model for light. Hertz noticed that the production of sparks from metallic plates was greatly altered by the presence of light, a characteristic that came to be known as the *photoelectric effect*. Classical electromagnetic wave theory could not account for this effect.

From diverse areas, all having in common an event involving the emission or absorption of light, the quantum of energy was spawned to establish a possible explanation of experimental results. The Planck assumption in blackbody radiation, Einstein's corpuscle of energy, Bohr's quantized atomic model, and the Compton effect, for example, all contributed to the evolution of a post-classical view incorporating the quantum of energy into a probablistic theory of quantum mechanics.

Although this new theory was created in order to understand particle behavior, the quantum of energy, which is a unit of radiation, was inextricably woven into its formulation. Yet, light received little attention from the Schrödinger wave equation approach to quantum theory. A description of light in quantized terms is necessary to attain a deeper understanding of the photon. For that purpose, the Schrödinger equation, cast in a more general form, will point the way towards the quantization of the classical electromagnetic field.

5.1 MAXWELL'S PROPAGATING FIELDS

A light field in free space is a superposition of electromagnetic waves characterized by amplitudes that can take on any finite value. The form of the waves is predicated on the fundamental properties of electric and magnetic fields, namely: Coulomb's law, the Biot–Savart law, Faraday's law, and Ampere's circuital law.

For a vacuum with no charges or currents present, these four laws are summarized by Maxwell's field equations as

$$\left(\frac{\partial}{\partial x}\hat{\mathbf{i}} + \frac{\partial}{\partial y}\hat{\mathbf{j}} + \frac{\partial}{\partial z}\hat{\mathbf{k}}\right)\cdot \mathbf{D} = \nabla \cdot \mathbf{D} = 0 \tag{5.1}$$

$$\nabla \cdot \mathbf{B} = 0 \tag{5.2}$$

$$\left(\frac{\partial \mathcal{E}_z}{\partial y} - \frac{\partial \mathcal{E}_y}{\partial z}\right)\hat{\mathbf{i}} + \left(\frac{\partial \mathcal{E}_x}{\partial z} - \frac{\partial \mathcal{E}_z}{\partial x}\right)\hat{\mathbf{j}} + \left(\frac{\partial \mathcal{E}_y}{\partial x} - \frac{\partial \mathcal{E}_x}{\partial y}\right)\hat{\mathbf{k}} = \nabla \times \mathcal{E} = -\frac{\partial \mathbf{B}}{\partial t} \tag{5.3}$$

and

$$\nabla \times \mathbf{H} = \frac{\partial \mathbf{D}}{\partial t} \tag{5.4}$$

with

$$\mathbf{B} = \mu_0 \mathbf{H}, \quad \mathbf{D} = \epsilon_0 \mathcal{E}, \quad \text{and} \quad \mu_0 \epsilon_0 = c^{-2}$$

The energy of the fields contained in a volume τ is given by

$$E = \frac{1}{2}\int_{\text{volume}} \left(\epsilon_0 \mathcal{E}^2 + \mu_0 \mathbf{H}^2\right) d\tau \tag{5.5}$$

Seeking a solution of Maxwell's equations for a one-dimensional field $\mathcal{E}_y$, allows that the field can be a function of x and t for the simplest choice of a nontrivial solution. In this case Eqs. (5.1) and (5.2) are automatically satisfied and Eqs. (5.3) and (5.4) become

$$\frac{\partial \mathcal{E}_y}{\partial x} = -\mu_0 \frac{\partial H_z}{\partial t} \tag{5.6}$$

and

$$\frac{\partial H_z}{\partial x} = -\epsilon_0 \frac{\partial \mathcal{E}_y}{\partial t} \tag{5.7}$$

showing that **H** must be in the z direction as a function of x and t. By differentiating Eq. (5.6) with respect to x and Eq. (5.7) with respect to t, we get a wave equation for $\mathcal{E}_y(x, t)$ in the form

$$\frac{\partial^2 \mathcal{E}_y}{\partial x^2} = \frac{1}{c^2} \frac{\partial^2 \mathcal{E}_y}{\partial t^2} \tag{5.8}$$

It is possible to eliminate the electric field in the above equations and arrive at a similar wave equation for the magnetic field, but since half of the energy is associated with the electric field and the other half with the magnetic field, it is necessary only to deal with one field. This allows a calculation of the total electromagnetic field energy as

$$E = \int_{\text{volume}} \epsilon_0 \mathcal{E}^2 \, d\tau \tag{5.9}$$

As we have seen, the one-dimensional wave equation admits a solution for any function in the form

$$f(x \mp ct) \tag{5.10a}$$

We have already invoked a particularly simple function which meets this requirement, namely

$$\mathcal{E}_y = \mathcal{E}_{0_y} \sin(kx - \omega t) \tag{5.10b}$$

which represents a plane wave with y displacements traveling in the positive x direction with a velocity $c = \omega/k$. For convenience we drop the subscript y and only deal with the electric field remembering however that the magnetic field

$$H_z = H_{0_z} \sin(kx - \omega t) \tag{5.11}$$

is in phase with the electric field vector but oriented in a perpendicular plane (Fig. 5.1).

It is convenient to impose a boundary condition on this solution which forces the wave to be cyclic within a distance L. This condition, which can be stated as

$$\mathcal{E}(x, t) = \mathcal{E}(x + L, t) \tag{5.12}$$

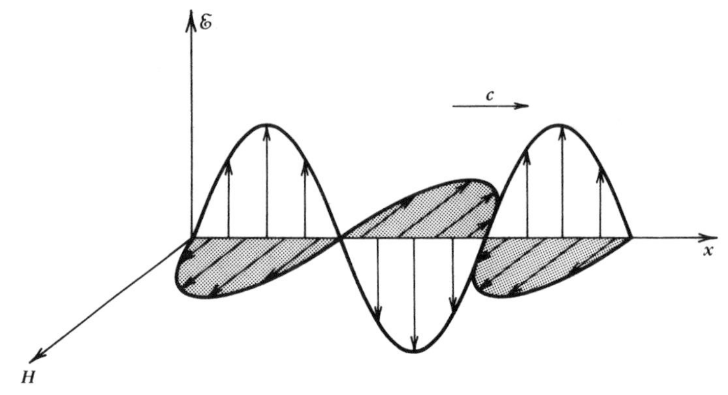

Figure 5.1. A traveling electromagnetic wave.

restricts the value of the wave numbers to an infinite discrete set

$$k = \frac{2\pi n}{L} \tag{5.13}$$

Since $\omega = kc$, a set of possible values ω_k is also generated. As a result, a more general solution to the wave equation is a linear sum of the traveling waves for each of the allowed values.

$$\mathcal{E} = \sum_k \mathcal{E}_{0_k} \sin(kx - \omega_k t) \tag{5.14}$$

Concentrating for now on just one component of this more general solution, we can calculate the energy associated with this component wave contained in the region L from

$$E_k = \int_0^L \epsilon_0 \mathcal{E}_k^2 \, dx \tag{5.15}$$

Although it is simple to carry out this calculation for the sinusoidal traveling wave, it is enlightening first to put the electric field solution in a slightly different but equivalent form, one in which the classical field amplitudes show a mathematical similarity to the variables of the harmonic oscillator.

5.2 NEW FIELD AMPLITUDES FOR OLD

Expanding the trigonometric solution for one component of the traveling wave (where the subscript k will be omitted), we have

$$\mathcal{E} = \mathcal{E}_0(\cos \omega t \sin kx - \sin \omega t \cos kx) \tag{5.16}$$

Allowing the definitions

$$q(t) = \frac{\mathcal{E}_0}{\omega} \cos \omega t \tag{5.17}$$

and

$$p(t) = -\mathcal{E}_0 \sin \omega t \tag{5.18}$$

we can write the electric field solution as

$$\mathcal{E} = \omega q(t) \sin kx + p(t) \cos kx \tag{5.19}$$

By definition, the functions $q(t)$ and $p(t)$ are 90° out of phase with one another and are related by the time derivative

$$p(t) = \dot{q}(t) \tag{5.20}$$

Other than being the time-dependent parts of our solution, the functions $p(t)$ and $q(t)$ are without physical significance at this point.

Returning to the energy calculation, Eq. (5.15), we have

$$E = \int_0^L \epsilon_0 [\omega q(t) \sin kx + p(t) \cos kx]^2 \, dx \tag{5.21}$$

which upon integration is

$$E = \tfrac{1}{2} \epsilon_0 L \left[\omega^2 q(t)^2 + p(t)^2 \right] \tag{5.22}$$

This can be tidied up by incorporating the constants in the functions

$$Q(t) = (\epsilon_0 L)^{1/2} q(t) \tag{5.23}$$

and

$$P(t) = (\epsilon_0 L)^{1/2} p(t) \tag{5.24}$$

Thus the energy of a single traveling plane wave in a region L is

$$E = \tfrac{1}{2} (\omega^2 Q^2 + P^2) \tag{5.25}$$

This result is extremely suggestive, since it has the same mathematical form as the energy of a simple harmonic oscillator.

The energy expression will prove to be an important guide subsequently for the quantization of the classical electromagnetic field. It will be expedient, therefore, to put it in a slightly different form from that just derived. Suppose we factor the quadratic $\omega^2 Q^2 + P^2$, and obtain a product of complex conjugates.

$$\omega^2 Q^2 + P^2 = (\omega Q + iP)(\omega Q - iP) \tag{5.26}$$

These combinations of the real time-dependent amplitudes P and Q are used to define complex amplitudes

$$a(t) = (2\hbar\omega)^{-1/2}(\omega Q + iP) \tag{5.27}$$

and

$$a^*(t) = (2\hbar\omega)^{-1/2}(\omega Q - iP) \tag{5.28}$$

Defined in this manner the complex amplitude function can be thought of as a rotating vector

$$a(t) = ae^{-i\omega t} \tag{5.29a}$$

where

$$a = \mathcal{E}_0\left(\frac{\epsilon_0 L}{2\hbar\omega}\right)^{1/2} \tag{5.29b}$$

making it start, at $t = 0$, on the real axis rotating clockwise in the complex plane. Its projections on the axes are the time-dependent amplitudes $(2\hbar\omega)^{-1/2}P(t)$ and $(2\hbar\omega)^{-1/2}\omega Q(t)$ (Fig. 5.2).

The energy expressed in terms of these newly defined amplitudes is

$$E = \hbar\omega a^*(t)a(t) = \hbar\omega a^2 \tag{5.30}$$

Although Planck's constant $2\pi\hbar = h$ now appears in the equation, it is only because of the arbitrary definition of $a(t)$. The energy is still the classical result, and has nothing as yet to do with the quantum theory.

Now that the energy of the wave is established in terms of the complex amplitudes, it would be helpful to reformulate the electric field solution. Using the identities

$$e^{ikx} = \cos kx + i \sin kx \tag{5.31}$$

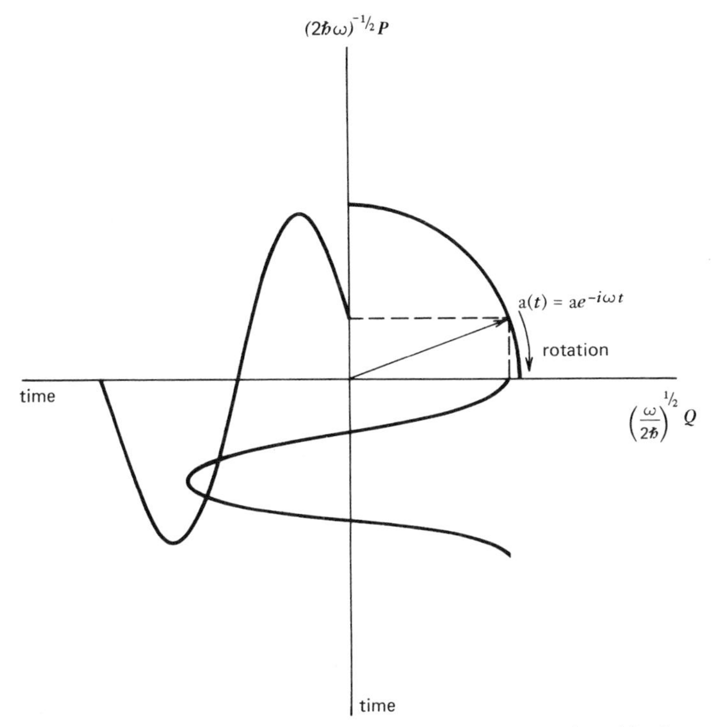

Figure 5.2. The complex amplitude $a(t)$ and its projections plotted in time.

and

$$e^{-ikx} = \cos kx - i \sin kx$$

we establish the electric field (5.19) as

$$\mathcal{E} = (\epsilon_0 L)^{-1/2}\left(\omega Q \frac{e^{ikx} - e^{-ikx}}{2i} + P\frac{e^{ikx} + e^{-ikx}}{2}\right) \tag{5.32}$$

With a little rearrangement we have

$$\mathcal{E} = \frac{i}{2}(\epsilon_0 L)^{-1/2}\left[(\omega Q - iP)e^{-ikx} - (\omega Q + iP)e^{ikx}\right] \tag{5.33}$$

whereupon the complex amplitudes may be introduced to give

$$\mathcal{E} = i\left(\frac{\hbar\omega}{2\epsilon_0 L}\right)^{1/2}\left[a^*(t)e^{-ikx} - a(t)e^{ikx}\right] \tag{5.34}$$

which is still nothing more than

$$\mathcal{E} = \mathcal{E}_0 \sin(kx - \omega t) \tag{5.35}$$

but in a form that will be quite useful.

The electric field is now expressed as the sum of a complex function $a(t)e^{ikx}$ and its complex conjugate $a^*(t)e^{-ikx}$ which we will call, respectively, the positive and negative parts of the field, namely,

$$\mathcal{E}^{(+)} = -i\left(\frac{\hbar\omega}{2\epsilon_0 L}\right)^{1/2} a(t)e^{ikx} \tag{5.36}$$

$$\mathcal{E}^{(-)} = i\left(\frac{\hbar\omega}{2\epsilon_0 L}\right)^{1/2} a^*(t)e^{-ikx} \tag{5.37}$$

Either of these complex electric fields contains the same information as the real field, and is related by

$$\mathcal{E} = 2\,\mathrm{Re}\,\mathcal{E}^{(+)} = 2\,\mathrm{Re}\,\mathcal{E}^{(-)} \tag{5.38}$$

The energy can be expressed in terms of these positive and negative parts as

$$E = 2\int_0^L \epsilon_0 \mathcal{E}^{(-)}\mathcal{E}^{(+)}\,dx \tag{5.39}$$

Let us review what we have done. Starting with a plane traveling wave solution with periodic boundary conditions, we expressed the electric field as

$$\mathcal{E} = \mathcal{E}_0 \sin(kx - \omega t) \tag{5.40}$$

and found the energy to be

$$E = \tfrac{1}{2}(\omega^2 Q^2 + P^2) \tag{5.41}$$

where

$$P = \dot{Q} = -(\epsilon_0 L)^{1/2}\mathcal{E}_0 \sin\omega t \tag{5.42}$$

Through a further definition of amplitudes

$$a(t) = (2\hbar\omega)^{-1/2}(\omega Q + iP) = ae^{-i\omega t} \tag{5.43}$$

and

$$a^*(t) = (2\hbar\omega)^{-1/2}(\omega Q - iP) = ae^{i\omega t} \tag{5.44}$$

we were able to formulate the real field as

$$\mathcal{E} = \mathcal{E}^{(-)} + \mathcal{E}^{(+)} \tag{5.45}$$

where

$$\mathcal{E}^{(+)} = (\mathcal{E}^{(-)})^* = -i\left(\frac{\hbar\omega}{2\epsilon_0 L}\right)^{1/2} a(t)e^{ikx} \tag{5.46}$$

with the field energy in the form

$$E = 2\int_0^L \epsilon_0 \mathcal{E}^{(-)}\mathcal{E}^{(+)}\,dx = \hbar\omega a^2 \tag{5.47}$$

With one further generalization, these results will appear in a form convenient for use in a quantum treatment of the field and its energy. Not just one but an infinite set of traveling waves

$$\mathcal{E} = \sum_k \mathcal{E}_{0_k}\sin(kx - \omega_k t) \tag{5.48}$$

one for each allowed value of $k = 2\pi n/L$, is the more general solution to the wave equation. In terms of the field and the energy we have finally the general solution

$$\mathcal{E}^{(+)} = -i\sum_k \left(\frac{\hbar\omega_k}{2\epsilon_0 L}\right)^{1/2} a_k(t)e^{ikx} \tag{5.49}$$

and

$$E = \tfrac{1}{2}\sum_k (\omega_k^2 Q_k^2 + P_k^2) \tag{5.50}$$

or

$$E = \sum_k \hbar\omega_k a_k^2 \tag{5.51}$$

This is as far as we need to go with the classical description of the field.

Every expression of the field energy shows it to be dependent on the square of an amplitude, a classical fact that will be challenged in subsequent applications.

5.3 BLACKBODY RADIATION DILEMMA

When heated, material bodies emit radiant energy, the nature of which depends on the temperature of the radiating body and is independent of the particular type of material being heated. For this reason it is expedient to define theoretically an ideal radiating surface that absorbs all energy incident upon it, reflecting none. All radiant energy emanating from this surface is emitted energy. This condition can be closely approximated (especially in the optical region) by coating a surface with finely granular, loosely packed carbon.

In an isothermal enclosure where no radiation can escape, all the radiation contained in the enclosure is emitted radiation, making its spectral composition equivalent to that emitted by an ideal black surface. Such a cavity is defined to be an ideal blackbody radiator, and can be approximated experimentally by designing an enclosure with a small hole through which the radiant energy can be observed.

The blackbody energy distribution for this experimental arrangement was measured as a function of wavelength at various temperatures of the cavity by Lummer and Pringsheim.‡ At any given temperature the intensity of radiant electromagnetic energy/wavelength is a function of the wavelength, peaking at some characteristic wavelength and trailing off to zero at both high and low wavelengths (Fig. 5.3).

The total intensity radiated by an ideal blackbody is expressed by

$$I(T) = \int_0^\infty I(\lambda, T)\, d\lambda \tag{5.52}$$

and is predicted by the Stefan–Boltzmann law to be

$$I(T) = \sigma T^4 \tag{5.53}$$

where T is absolute temperature and $\sigma = 5.67 \times 10^{-8}\ \mathrm{W}/m^2\ \mathrm{K}^4$.

Wien proposed an empirical equation to express the intensity per given wavelength (called the *monochromatic emissive power*) as

$$I(\lambda, T) = \frac{a_1 e^{-a_2/\lambda T}}{\lambda^5} \tag{5.54}$$

‡Lummer, O. and E. Pringsheim, *Verhandlungen der Deutschen physikalischen Gesellschaft* **1**, 23, 215 (1899); **2**, 163 (1900).

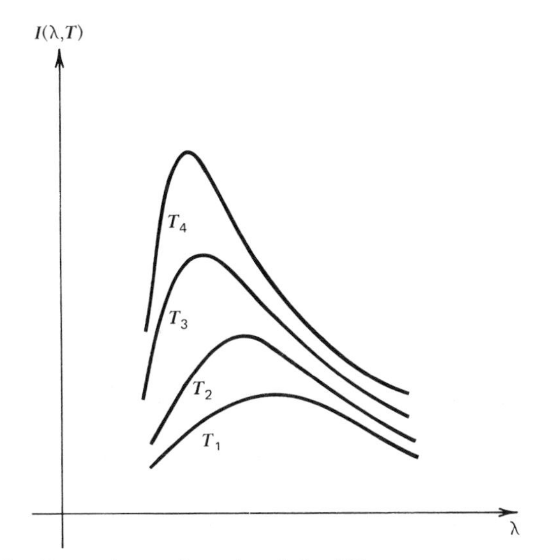

Figure 5.3. Energy/spectral wavelength for different temperatures, $T_4 > T_1$.

where a_1 and a_2 are evaluated by fitting Wien's equation to the experimental results. Not only is this result unsatisfying since it provides no physical argument by which to evaluate the constants, but more importantly, it fits the experimental results closely only at wavelengths shorter than the wavelength of maximum emission (Fig. 5.4).

By contrast, Rayleigh and Jeans developed a physical approach to derive a relation for the monochromatic emissive power. Their results fit the experimental data at long wavelengths, but predict an "ultraviolet catastrophe" at shorter wavelengths, that is, in direct contrast with Wien's

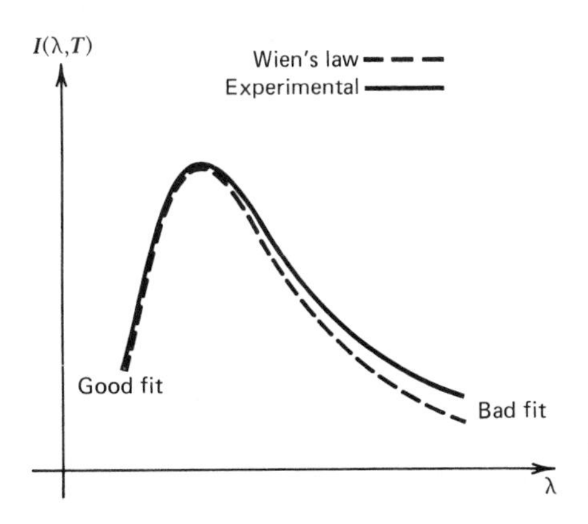

Figure 5.4. Comparison of Wien's law with experimental energy distribution results.

law, the emissive power increased without bound for smaller wavelengths or higher frequencies. Under their reasoning, Rayleigh and Jeans envisioned a model blackbody radiator as a large number of radiating charged particles undergoing simple harmonic motion. Their energy is proportional to the square of the amplitudes. The average energy assigned to each mode of the oscillators by the *law of equipartition of energy* is KT (where K is the Boltzmann constant) which represents $\frac{1}{2}KT$ for each of two degrees of freedom, one for kinetic energy and the other for potential energy. By multiplying the average energy of each mode of oscillation by the density of the number of modes/wavelength at a particular wavelength, they obtained the radiant energy density/wavelength for the blackbody radiator. To understand their approach we shall look at a calculation of the number of modes of oscillation.

Assuming for convenience a cubical enclosure with perfectly reflecting walls, the electric field parallel to the wall must vanish. This condition generates standing waves with an appropriate set of allowed wave numbers $k_n = n\pi/L$ where $n = 1, 2, \ldots,$ and L is the dimension of the cavity. If we generalize the result to three dimensions, the allowed modes correspond to wave numbers given by

$$k_n = \frac{\pi}{L}\left(n_x^2 + n_y^2 + n_z^2\right)^{1/2} = \frac{2\pi}{\lambda_n} \tag{5.55}$$

where $n = (n_x^2 + n_y^2 + n_z^2)^{1/2}$.

Visualizing a lattice composed of points representing each possible set of integers (n_x, n_y, and n_z), we can calculate the number of modes/unit volume at a given wavelength in lattice space (Fig. 5.5). Concentrating on one octant which represents the positive integers and understanding that the number of points is proportional to the volume in lattice space (especially where L is large making the points densely packed), we determine that the number of modes falling within concentric spheres of radius n and $n + dn$ is

$$dN = \left(\tfrac{1}{8}\right)4\pi n^2\,dn \tag{5.56}$$

If we use the radius in lattice space from Eq. (5.55), namely,

$$n = \frac{2L}{\lambda_n} \tag{5.57}$$

and differentiate with respect to wavelength

$$dn = \frac{2L}{\lambda_n^2}\,d\lambda_n \tag{5.58}$$

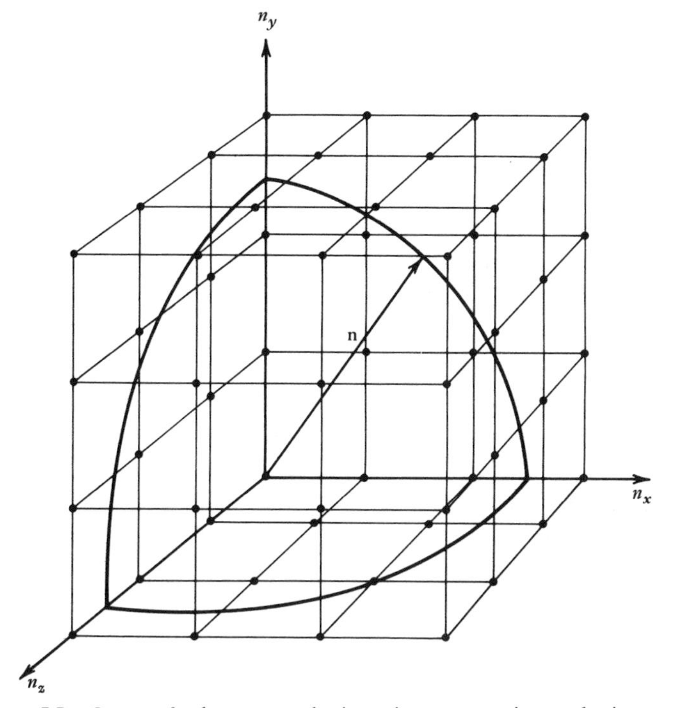

Figure 5.5. Octant of sphere among lattice points representing modes in a cavity.

where the negative sign from the differentiation is ignored since dn must be positive. The number of modes becomes

$$dN = \left(\frac{1}{8}\right)4\pi\frac{(2L)^3}{\lambda_n^4}d\lambda_n(2) \tag{5.59}$$

where the result is multiplied by 2 since each standing wave is really two traveling waves moving in opposite directions. Finally the density of the number of modes/wavelength is

$$\frac{1}{L^3}\frac{dN}{d\lambda_n} = \frac{8\pi}{\lambda_n^4} \tag{5.60}$$

making the energy density/wavelength in the Rayleigh–Jeans calculation become

$$U(\lambda, T) = \frac{8\pi}{\lambda^4}KT \tag{5.61}$$

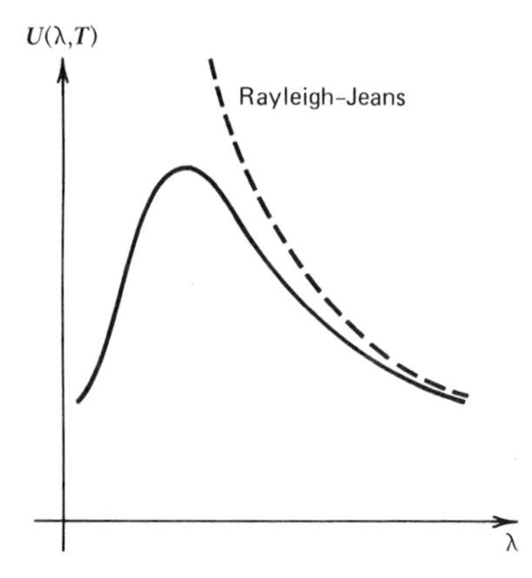

Figure 5.6. Comparison of Rayleigh–Jeans calculation with experimental results.

While the result matches well the experimental values for long wavelengths, it diverges as λ approaches zero, predicting runaway energies for high frequencies, hence the "catastrophe" (Fig. 5.6). This failure of classical physics set the stage for Max Planck to introduce a radical assumption which gave rise to the quantum theory.

Sceptical of the equipartition of energy and finding a manipulation of Wien's empirical equation, Planck was led to suspect that the energy of an oscillator was related to its frequency, an assumption breaking drastically with the results of classical physics, whereby the energy of an oscillator is related to the square of its amplitude. Moreover, he assumed oscillators to have discrete energy states given by

$$E_n = nhf \tag{5.62}$$

where n is an integer and the value of h was found by fitting his results with the experimental data.

Under this assumption, an oscillator can absorb or emit a quantum of energy

$$\Delta E = hf \tag{5.63}$$

by jumping from one quantized state to another. Normally an oscillator is found in its ground state (state of lowest energy) unless agitated by an absorption of energy. A large collection of thermally excited oscillators will distribute among the allowed energy levels according to the Maxwell–Boltzmann distribution function

$$N(nhf) = N_0 e^{-nhf/KT} \tag{5.64}$$

The average energy of oscillators distributed over the allowed quantum states becomes

$$\langle E \rangle = \frac{\sum_{n=0}^{\infty} N(nhf) E_n}{\sum_{n=0}^{\infty} N(nhf)} = \frac{\sum_{n=0}^{\infty} N_0 e^{-nhf/KT} nhf}{\sum_{n=0}^{\infty} N_0 e^{-nhf/KT}} \tag{5.65}$$

$$= hfx \frac{(1 + 2x + 3x^2 + \cdots)}{(1 + x + x^2 + \cdots)} \tag{5.66}$$

where $x = e^{-hf/KT}$, giving

$$\langle E \rangle = hfe^{-hf/KT} \frac{(1-x)^{-2}}{(1-x)^{-1}} = \frac{hf}{e^{hf/KT} - 1} \tag{5.67}$$

Using Eq. (5.60) for the density of modes/wavelength, Planck's radiation law predicts the energy density/wavelength to be

$$U(\lambda, T) = \frac{8\pi hc}{\lambda^5} \left(e^{hc/\lambda KT} - 1\right)^{-1} \tag{5.68}$$

in fine agreement with experimental observations. This reduces to the Rayleigh–Jeans prediction, Eq. (5.61), in the limit of large wavelengths and the Wien law at short wavelengths.

5.4 PHOTOELECTRIC EFFECT AND EINSTEIN'S MASSLESS PARTICLES

When monochromatic light of high enough frequency falls on a metal, electrons are ejected. A change in the frequency of radiation changes the maximum kinetic energy of the electrons (Fig. 5.7), while a change in the intensity of the light beam does not effect the energy of the electrons. Varying the intensity does change the number of photoelectrons ejected per second, but not their maximum kinetic energy (Fig. 5.8). As long as we remain above the threshold frequency f_0, photoelectrons are emitted practically instantaneously ($< 10^{-9}$ s) even for exceedingly low intensity levels.

These experemental facts defy the classical explanation according to which the intensity, not the frequency, should determine the energy absorbed by the electron. At exceedingly low intensities it should take a long time for an electron to accumulate enough energy to surpass its lowest binding energy (*the work function* W_0) and leave the metal.

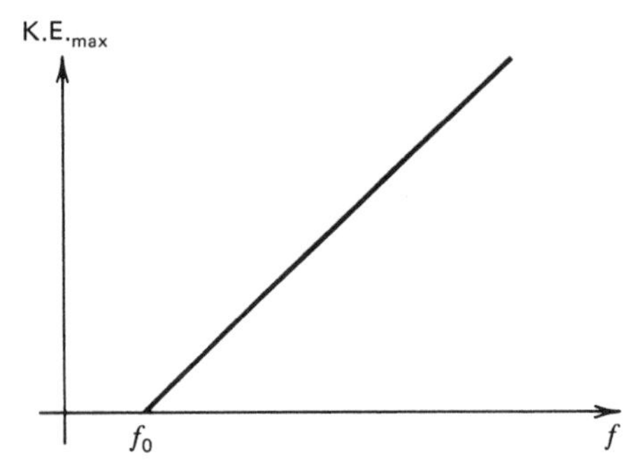

Figure 5.7. Variation of maximum kinetic energy of photoelectrons with frequency.

Albert Einstein postulated, that a quantum of energy $E = hf$ was absorbed from the incident radiation‡ by an electron allowing an energy transaction expressed by

$$hf = W_0 + \tfrac{1}{2}mv_{\text{max}}^2 \tag{5.69}$$

This quantum of energy provides the energy to overcome the minimum binding energy W_0 and then supplies any excess energy to the kinetic energy of the photoelectron. This explained the minimum threshold frequency necessary to observe the photoelectric effect:

$$f_0 = \frac{W_0}{h} \tag{5.70}$$

as well as showing the maximum kinetic energy of the photoelectron to be dependent linearly upon the frequency of the incident radiation. Furthermore, the rate/unit area at which photons arrive at the surface is related to

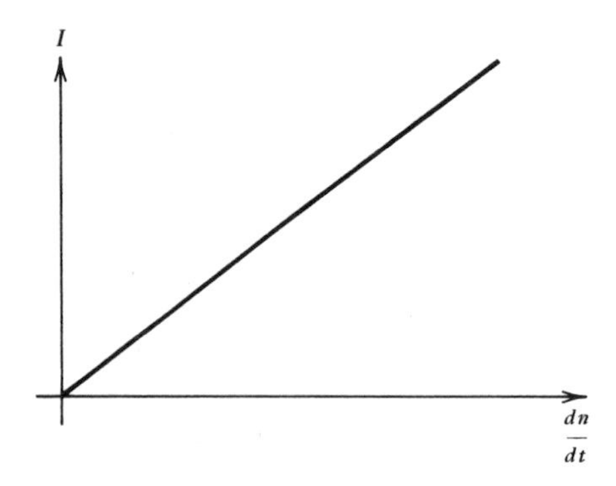

Figure 5.8. Variation of intensity with photoelectron rate.

‡Einstein, Albert. *Ann. Phys.* **17**, 132 (1905).

the intensity of the radiation by

$$I = \frac{1}{A}\frac{dn}{dt}hf \tag{5.71}$$

Thus, Planck's quantum became Einstein's photon, and somehow electromagnetic waves, so firmly established by Maxwell's equations and Young's experiment, appear paradoxically as a stream of photons whose energy is related to the wave frequency. The resolution of these conflicting models for light, namely, wave versus particle, will occupy a good part of the rest of this text and become the basis for a deeper understanding of light.

In the same year as the appearance of the photoelectric paper, Einstein published his special theory of relativity, one result of which was the variation of rest mass with velocity as expressed by‡

$$m = m_0\left(1 - \frac{v^2}{c^2}\right)^{-1/2} \tag{5.72}$$

where m_0 is the rest mass of a particle and c, the speed of light. With the energy of a mass m expressed as

$$E = mc^2 \tag{5.73}$$

a rearrangement of Eq. (5.72) yields

$$E^2 - p^2c^2 = E_0^2 \tag{5.74}$$

where p is the momentum of the moving particle and E_0 is the rest energy, $E_0 = m_0c^2$.

In the case of the massless particle of the photoelectric effect, we gain the following expression for the photon by setting $m_0 = 0$:

$$E = pc \tag{5.75}$$

which indicates that the photon, as a particle with zero rest energy, has both energy and momentum. Combining this allowance of relativity theory with the photoelectric assumption for the photon energy we obtain

$$E = hf = pc \tag{5.76}$$

‡Einstein, Albert. *Ann. Phys.* **17**, 89 (1905).

making the photon momentum

$$p = \frac{h}{\lambda} \tag{5.77}$$

thus advancing the "particle" aspects of the photon.

5.5 MATTER WAVES AND THE UNCERTAINTY PRINCIPLE

Assuming a symmetry between mass and energy, Prince Louis de Broglie reasoned that if wavelike radiant energy can behave as a particle, then ordinary matter, that is, the particle, could have wavelike characteristics. Guided by the energy and momentum relations for a photon, de Broglie proposed that a mass particle is accompanied by a wave whose frequency and wavelength are determined by

$$f = \frac{E}{h} \quad \text{and} \quad \lambda = \frac{h}{p} \tag{5.78}$$

Unlike electromagnetic waves, however, whose phase speed in a vacuum is that of light, the phase speed of de Broglie's matter waves become

$$u = \lambda f = \frac{E}{p} \tag{5.79}$$

which for a relativistic particle of energy and momentum $E = mc^2$ and $p = mv$ yields

$$u = \frac{mc^2}{mv} = \frac{c^2}{v} \tag{5.80}$$

and for a free nonrelativistic particle of energy $E = p^2/2m$ and momentum $p = mv$ yields

$$u = \frac{v}{2} \tag{5.81}$$

In neither case does the phase wave follow the particle which is moving with a speed v. As a matter of fact, in the first case the wave is moving faster than the speed of light,[‡] and in the second case it moves at half the

[‡] There is no problem with phase speeds greater than that of light since energy is transported at group velocities, thereby not violating the theory of relativity.

particle speed. Thus this phase wave cannot have a physical reality of its own but nonetheless could be responsible for the interference effects observed for particles. It is tempting to associate a wave packet with the particle; not only would that allow us to relate the group velocity of the packet to the particle velocity, but also to picture the energy of the wave confined in a narrow region approximately in the vicinity of the particle.

However, care must be taken. These matter waves are *not* ordinary waves and one cannot think of the particle as actually being a wave packet. For one thing we cannot ascribe a mass to a wave packet; for another the width of a wave packet is bound in a fundamental way to the spread in values of the wavelengths for the phase waves that compose it. This relation for classical wave packets takes the form

$$\Delta x\, \Delta k \simeq 1 \tag{5.82}$$

Applying the de Broglie criterion for wavelengths of matter waves, namely, $\Delta p = \hbar\, \Delta k$, leads to a profound utterance on nature.

$$\Delta x\, \Delta p \simeq \hbar \tag{5.83}$$

The classical particle, which we have always thought of as having a precise, definite position and momentum capable of being measured simultaneously at any time, can no longer be conceived in this manner if a particle is to be represented by matter waves. A simultaneous diffuseness in both position and momentum must exist whose product is of the order of Planck's constant. This does not mean that we cannot have a particle of rather precise position or momentum, just that we cannot have them together. A particle represented by a very narrow packet can be well localized with an exceedingly small Δx but the packet will be composed of a broad set of phase waves whose Fourier transform will require a large spread in momentum values Δp. Conversely, a particle of definite momentum (or wavelength), must be represented by a matter wave with a broad spatial extent.‡ This uncertainty principle is perhaps best illustrated by an interference example for particles that is quite similar in structure to Young's experiment for waves.

Let us invoke the Young apparatus of a barrier with two slits, but instead of monochromatic light, we shall allow a beam of electrons having a definite momentum to impinge upon the slits (Fig. 5.9*a*). After many electrons have

‡Goldin, E., J. Bregman and D. Scarl. *The Packet of an Uncertain Guassian*, International Film Bureau, Inc., Chicago, 1982.

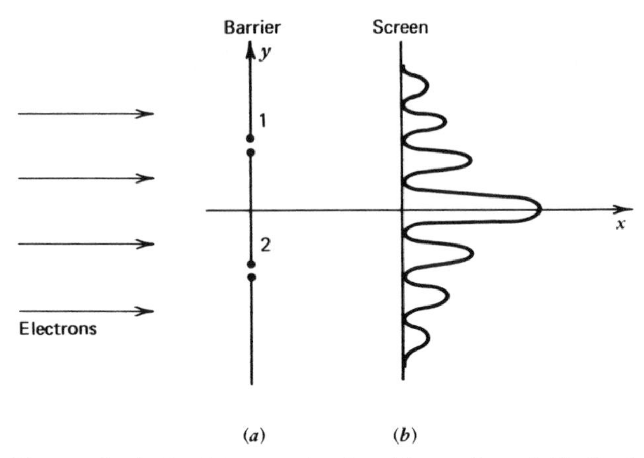

Figure 5.9. Electrons impinging (*a*) upon a barrier with two slits and (*b*) distributing upon a detecting screen.

passed through the slits the pattern of distribution of electrons at the screen will be similar to the interference pattern of waves (Fig. 5.9*b*), providing very strong evidence that the electrons are undergoing a wavelike interference. Even if one electron at a time is projected towards the slits, after many electrons have been detected at the screen, the same pattern occurs. Unlike waves which interfere with one another, the single electron, strangely enough, appears to interfere with itself. It is almost as though the electron is capable of going through both slits at the same time. It surely means that we cannot tell which slit it went through, and any attempt to do so, say, by closing one slit or the other, destroys the interference pattern (Fig. 5.10*a* or 5.10*b*).

With either slit closed the position of the particle is determined to within a narrow Δy (the width of the slit) requiring therefore a sizeable uncertainty in the momentum Δp_y. The beam is no longer monochromatic, making interference impossible, and allowing the electron to turn up at a broad set of positions on the screen. Just to round out the picture, if the single slit of Fig. 5.10*a* is made broader, that is, Δy is larger, then Δp_y will become smaller giving a narrower distribution of electrons, as in Fig. 5.11.

As in Young's experiment, if matter is represented by some sort of wave ψ, then the electrons reaching the screen with both slits open superimpose their wave states from each slit, so that the composite wave at the screen is

$$\psi = \psi_1 + \psi_2 \tag{5.84}$$

Following the idea that the intensity of a real wave is related to its square, we assume that the square of a matter wave represents the relative distribu-

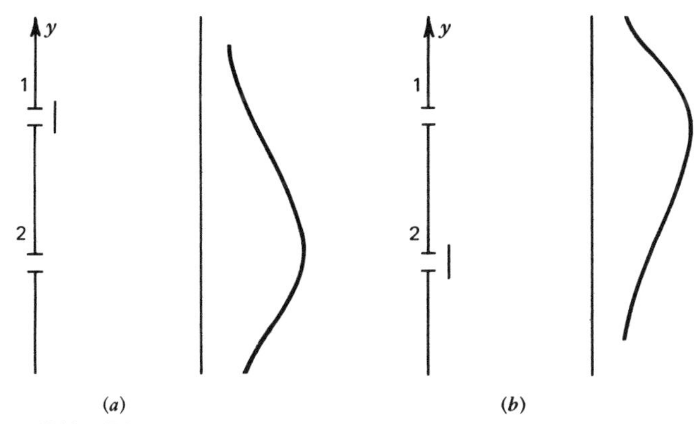

Figure 5.10. Distribution of electrons with (a) one, or (b) the other slit closed.

tion of electrons at the screen in which case

$$|\psi_1 + \psi_2|^2 = |\psi_1|^2 + |\psi_2|^2 + \psi_1\psi_2^* + \psi_2\psi_1^* \tag{5.85}$$

where the complex conjugate is used to allow that matter waves can be complex wave functions. The last two terms of Eq. (5.85) give rise to the interference pattern observed for the two slits, whereas, $|\psi_1|^2$ and $|\psi_2|^2$ represent the distribution of electrons for either slit open only one at a time. Thus, the evidence is strong for the introduction of matter waves as proposed by de Broglie. It remains for us to find a formulation of matter waves in terms of the physical properties of any system.

5.6 SCHRÖDINGER'S QUANTUM FORMULATION

In an attempt to assimilate these ideas, Erwin Schrödinger embodied de Broglie's waves with a wave equation. Using Schrödinger's approach for dealing with a nonrelativistic expression for energy and momentum of

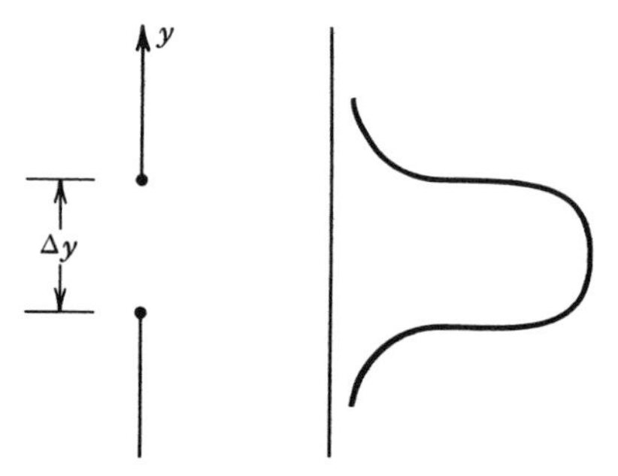

Figure 5.11. Electron distribution for a broader slit.

matter, we have

$$E = hf = \frac{p^2}{2m} + V \tag{5.86}$$

where V is the potential energy of the system and the momentum is the de Broglie expression

$$p = \frac{h}{\lambda} = mv \tag{5.87}$$

Thus, Schrödinger sought waves of frequency and wavelength

$$f = \frac{p^2}{2mh} + \frac{V}{h} = \frac{E}{h} \tag{5.88a}$$

and

$$\lambda = \frac{h}{p} \tag{5.88b}$$

In analogy with a plane traveling wave where

$$f(x, t) = A \sin 2\pi\left(ft - \frac{x}{\lambda}\right) \tag{5.89}$$

a matter wave might take a similar form. But in order to form a differential wave equation and satisfy Eq. (5.86) at the same time we will need a first derivative in time to obtain the energy E and a second derivative in the spatial coordinate x to obtain p^2. This rules out the use of the trigonometric function which changes form on differentiation, but points to the complex function

$$\Psi(x, t) = Ae^{-i(2\pi/h)(Et-px)} \tag{5.90}$$

which maintains the same form. Considering individually then the terms

$$-\frac{h}{2\pi i}\frac{\partial\Psi}{\partial t} = E\Psi \tag{5.91a}$$

$$-\frac{1}{2m}\frac{h^2}{(2\pi)^2}\frac{\partial^2\Psi}{\partial x^2} = \frac{p^2}{2m}\Psi \tag{5.91b}$$

and

$$V\Psi = V\Psi \tag{5.91c}$$

the conservation of energy $E = p^2/2m + V$ dictates that the differential equation takes the form

$$\left[-\frac{h}{2\pi i}\frac{\partial}{\partial t}\right]\Psi(x,t) = \left[-\frac{h^2}{2m(2\pi)^2}\frac{\partial^2}{\partial x^2} + V\right]\Psi(x,t) \qquad (5.92)$$

which is the time-dependent one-dimensional Schrödinger wave equation. The de Broglie matter wave, referred to as the wave function $\Psi(x, t)$, can be solved for, once the potential energy of the particular physical system is specified.

The differential equation is unusual since it contains the imaginary number i and has solutions which are complex functions. The wave function Ψ cannot have a direct physical interpretation. As a matter of fact, if E is negative, as might occur with a strongly negative potential energy, the particle might still travel in the same direction, but the wave will be reversed.

The burden of understanding the significance of matter waves falls on an interpretation of Ψ. Here we take a clue from the observation of light waves. It is not the electromagnetic field but rather the intensity that is observed in a detection of light. Thus, if Ψ represents the state of a particle, we shall take the square, or since Ψ is a complex function, the product with its complex conjugate

$$\rho(x,t)\,dx = \Psi^*\Psi\,dx \qquad (5.93)$$

as the relative intensity or probability of observing the particle in an interval dx. The total probability of finding the particle somewhere becomes

$$P = \int_{-\infty}^{\infty} \Psi^*\Psi\,dx = 1 \qquad (5.94)$$

where $\rho(x, t) = \Psi^*\Psi$ takes on the role of a probability density. The condition defined by Eq. (5.94) provides a way of evaluating the coefficient or amplitude of the matter wave, a process referred to as *normalization*. It also makes demands on the finiteness of Ψ to be an acceptable wave function.

When the Schrödinger equation is generalized to three dimensions the momentum becomes $p^2 = p_x^2 + p_y^2 + p_z^2$. From Eq. (5.92b) we have already associated the component of the momentum with the operation

$$p_x^2 \rightarrow -\left(\frac{h}{2\pi}\right)^2\frac{\partial^2}{\partial x^2} = \left(i\frac{h}{2\pi}\frac{\partial}{\partial x}\right)\left(i\frac{h}{2\pi}\frac{\partial}{\partial x}\right) \qquad (5.95a)$$

which suggests the definition of a mathematical term as a momentum operator

$$\mathbf{p}_{x_{\text{op}}} = i\frac{h}{2\pi}\frac{\partial}{\partial x} \tag{5.95b}$$

Using this as a guide we ha· e the operation

$$\mathbf{p}_{\text{op}}^2 = -\left(\frac{h}{2\pi}\right)^2\left(\frac{\partial^2}{\partial x^2} + \frac{\partial^2}{\partial y^2} + \frac{\partial^2}{\partial z^2}\right) = \left(\frac{ih}{2\pi}\right)^2 \nabla^2 \tag{5.96}$$

in terms of which Schrödinger's equation becomes

$$\left[-\frac{\hbar^2}{2m}\nabla^2 + V(\mathbf{r}, t)\right]\Psi(\mathbf{r}, t) = \left[i\hbar\frac{\partial}{\partial t}\right]\Psi(\mathbf{r}, t) \tag{5.97}$$

where $\hbar$ has been used for $h/2\pi$. If the potential energy is independent of time, there is no explicit time dependence in the differential equation. Ψ can then be expressed as the product of two separate functions of the variables in the form

$$\Psi(\mathbf{r}, t) = \psi(\mathbf{r}) \cdot e^{-(i/\hbar)Et} \tag{5.98}$$

where $\Psi(\mathbf{r})$ is a stationary or time-independent wave function. A substitution of Eq. (5.98) in the differential equation (5.97) yields

$$\left[-\frac{\hbar^2}{2m}\nabla^2 + V(\mathbf{r})\right]\psi(\mathbf{r}) = E\psi(\mathbf{r}) \tag{5.99}$$

as the time-independent Schrödinger equation with E representing the total energy of the system.

In this form Schrödinger's wave mechanics can be applied to a variety of physical systems hosting a mass particle moving under a potential energy $V(\mathbf{r})$, such as a harmonic oscillator, a particle in a box, or the electron in a hydrogen atom. The purpose of introducing it here is to understand the application of quantum theory to electromagnetic fields and its paradoxical counterpart, the photon. Although the quantum of energy is already built into the theory in the appearance of h, and with E/h playing the role of frequency f in the wave function, we will have to cast the wave equation in a more universal form to see its application to massless entities.

The Schrödinger equation grew out of an approach to find a differential equation for matter waves describing particles. The wave equation is neither unique nor complete since we considered only the kinetic and potential energy of a nonrelativistic mechanical particle. Going back to the process of constructing a wave equation, we desired a differential equation with wave functions that would preserve the conservation of energy $E = p^2/2m + V$. In the process we found from Eqs. (5.91) and (5.95) that it is possible to associate a mathematical operation with each energy term. By formalizing the concept that was suggested in Eq. (5.95), we can define mathematical operators for the momentum and potential energy as

$$\mathbf{p} = i\hbar\nabla \tag{5.100a}$$

and

$$\mathbf{V}(\mathbf{r}) = V(\mathbf{r}) \tag{5.100b}$$

from which an operator for the total energy of the system, the Hamiltonian, becomes

$$\mathbf{H} = \frac{\mathbf{p}^2}{2m} + \mathbf{V}(\mathbf{r}) \tag{5.101}$$

Using this operator in the stationary Schrödinger equation (5.99) we have succinctly

$$\mathbf{H}\psi = E\psi \tag{5.102}$$

which is still the wave equation for matter waves if the Hamiltonian **H** is given by Eq. (5.101), but it is now in a more powerful form if we allow **H** to be constructed more generally. With classical variables and the structure of classical energy terms as a guide it is possible to construct Hamiltonian operators for more complex physical structures involving, for example, two or more interacting particles, rotational properties, interaction with electric and magnetic fields, and even systems of massless fields. In each case the construction of the Hamiltonian is generally guided by its classical counterpart. But in some cases, such as particle spin, no classical analogue exists, leaving the creation of the operator to a well-calculated mathematical guess, the verity of which is tested against experimental results.

In particular, we are interested in the ability to quantize electromagnetic fields. As we saw in the beginning of this chapter, the energy of a classical wave can be expressed in terms of its electric and magnetic fields. A redefinition of the variables of the field enabled us to define a single variable $a(t)$ for which both the fields and their energy could be expressed. It is this variable that will ultimately become a candidate for defining an operator and a Hamiltonian of the electromagnetic field. But because of the structural similarity of the field energy to that of a harmonic oscillator, we will first consider a solution of the quantized harmonic oscillator as a guide to quantizing the electromagnetic field.

Although the harmonic oscillator wave functions can be found through the Schrödinger wave equation (5.97), their solution is mathematically clumsy and is dealt with in many texts on quantum theory.[‡] Instead we will use the more general formulation $\mathbf{H}\psi = E\psi$ which we must use in any case, for quantizing the electromagnetic field.

A few remarks about the wave function ψ will help facilitate the task. What we wish to predict are the energies and values of physical properties of a system. The wave function itself is not useful, except insofar as it is necessary to evaluate probability densities $\psi^*\psi$ and the average values of operators for a state of the system. The average value of the Hamiltonian, for instance, is defined by

$$\langle \mathbf{H} \rangle = \psi^* \mathbf{H} \psi = \psi^* E \psi = E \psi^* \psi \tag{5.103}$$

Thus, if we know the operator rules for an operator acting on ψ states and have a means of expressing $\psi^*\psi$ as a real number, ψ itself need not be an ordinary or explicit mathematical function.

Finally, not just the Hamiltonian but any operator acting on a state ψ will produce a number a such that

$$\mathbf{A}_{\text{op}} \psi = a\psi \tag{5.104}$$

This, then is the heart of the process for the quantization of a system, with the precaution that not all values a of the operator are either real or physically meaningful—that depends on expedient (or possibly even creative) conceptualization and construction of the operator, the final arbiter being the description of reality.

REFERENCES

Anderson, E. E. *Modern Physics and Quantum Mechanics*, Saunders, Philadelphia, 1971.

Beiser, A. *Concepts of Modern Physics*, McGraw-Hill, New York, 1981.

Kim, S. K. and E. N. Strait. *Modern Physics for Scientists and Engineers*, Macmillan, New York, 1978.

[‡]See, for example, Powell, J. L. and B. Crasemann. *Quantum Mechanics*, Addison-Wesley, Reading, Mass., 1965, pp. 127–139.

Lindsay, R. B. and H. Margenau. *Foundations of Physics*, Dover, New York, 1957.

Richtmyer, F. K., E. H. Kennard, and T. Lauritsen. *Introduction to Modern Physics*, McGraw-Hill, New York, 1955.

PROBLEMS

5.1. For the scalar wave equation

$$\frac{\partial^2 V}{\partial x^2} = \frac{1}{c^2}\frac{\partial^2 V}{\partial t^2}$$

using $\xi = x - ct$ and $\eta = x + ct$ show that a general solution is $V = f(\xi) + g(\eta)$.

5.2. Evaluate the integral in Eq. (5.21).

5.3. Use the definition of the complex amplitudes a and a^* to show that Eq. (5.34) reduces to $\mathcal{E} = \mathcal{E}_0 \sin(kx - \omega t)$.

5.4. Explain the relation between the energy density/wavelength $U(\lambda, T)$ and the intensity/wavelength $I(\lambda, T)$.

5.5. Show that Planck's radiation law reduces to the Rayleigh–Jeans law in the limit of large wavelengths.

5.6. Using a short wavelength approximation, show that Planck's law reduces to Wien's law.

5.7. Show that the Planck radiation law can be written as

$$U(f, T) = \frac{8\pi h f^3}{c^3}\left(e^{hf/KT} - 1\right)^{-1}$$

5.8. The total energy density for the spectral distribution in blackbody radiation is Stefan's law and is found from

$$U(T) = \int_0^\infty U(f, T)\, df = \frac{4}{c}\sigma T^4$$

Perform the integration and show

$$\sigma = \frac{2\pi^5 K^4}{15 h^3 c^2}$$

5.9. Derive the result $E^2 - p^2c^2 = E_0^2$ from Eqs. (5.72) and (5.73) using the relativistic definition of momentum.

5.10. Using the de Broglie relations, show that the group velocity for a particle wave is given by

$$v = \frac{1}{\hbar}\frac{dE}{dk}$$

5.11. What is the uncertainty in momentum for an electron confined in an atom of diameter d. Let the uncertainty in position Δx be that diameter. What is the corresponding uncertainty in energy?

5.12. Show that the uncertainty relation for a free particle can be written as

$$\frac{2\pi \Delta\lambda \, \Delta x}{\lambda^2} \simeq 1$$

5.13. Using $\Psi(\mathbf{r}, t) = \psi(\mathbf{r}) \cdot \Phi(t)$ in the time-dependent Schrödinger equation obtain $\Phi(t) = ce^{-(i/\hbar)Et}$ and the time-independent Eq. (5.99).

5.14. If $\psi_1(x, t)$ and $\psi_2(x, t)$ are solutions to Schrödinger's equation, show that $\psi(x, t) = c_1\psi_1 + c_2\psi_2$ is also a solution, where the c's are arbitrary constants making $\psi(x, t)$ a linear superposition of the particular solutions.

5.15. Obtain a solution of Schrödinger's equation for a free particle where the potential energy, $V = 0$. Find the probability density for this particle.

5.16. Write Schrödinger's equation in one-dimension for a particle moving in a potential $V = \frac{1}{2}Kx^2$. Can you find one solution to this equation? *Hint*: Try, $\psi = Ae^{-ax^2}$. What is the corresponding energy to this solution?

6

A Quantum Recipe

One approach to quantum theory is through Schrödinger's second-order differential equation; the solutions yield predicted values for the physical properties of a system and their associated wave functions. Following an interpretation by Max Born that the square of a wave function is a probability distribution, we imagine these waves as probability field amplitudes of the whereabouts and activities of the physical system (Fig. 6.1).

A less picturesque but mathematically more advantageous approach is found in the quantum formulation of P. A. M. Dirac. With it we may trade the mathematics of differential calculus for the algebra of operators and state vectors, tools which are especially suitable for dealing with the harmonic oscillator and the subsequent quantization of the electromagnetic field. A simplified version of some of the Dirac formulation, therefore, is presented for the task at hand.

6.1 ELEMENTS OF THE THEORY

If we make a refined measurement of a physical property **A** on some system, we obtain in this single experiment some numerical result a_n. If we continue to make a large number of identical measurements on similarly prepared systems, we find eventually that not only does each set of values a_n appear as the result of the measurements, but also that each of the a_n's appears more or less often with respect to the others. The experimental information obtained is a set of values a_n and a relative frequency of appearance p_n, which is interpreted as the probability of finding a particular a_n in a single measurement (Fig. 6.2).

In view of these observations, quantum theory should predict a set of values a_n which have a probability of occurrence p_n, where each value a_n is associated with the behavior or operation of a physical property **A** for a system in some originally prepared condition or state. Our task, then, is to assign mathematical entities to states of the system and operations of physical properties, and with them, to study the assumptions and postulates

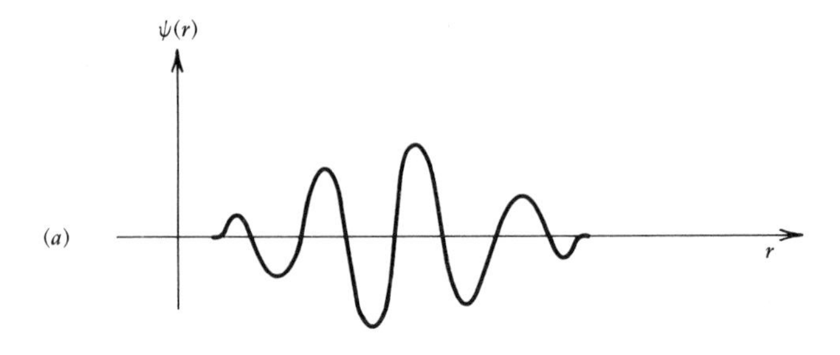

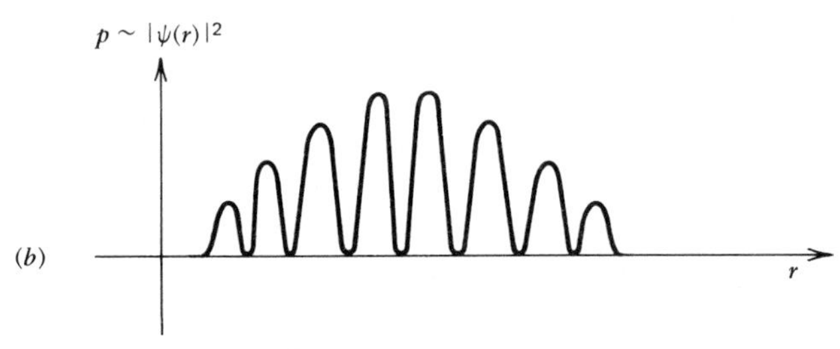

Figure 6.1. A rendition of (a) a wave function $\psi(r)$, and (b) its relation to probability interpretation p.

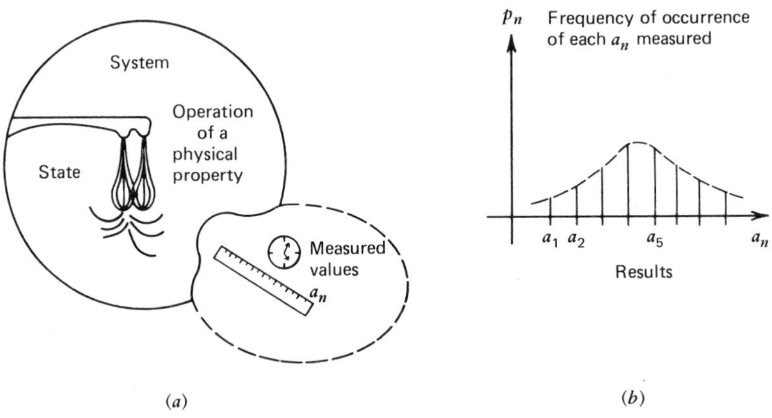

Figure 6.2. (a) Making repeated measurements on an identically prepared system: (b) a possible probability distribution of the measured results.

that lead to the desired physical results. In the theory of classical mechanics when values, for example, of position and momentum are desired, we find it necessary to introduce ideas such as mass and force, uncover postulates that relate the quantities, and develop the mathematics necessary to solve for the desired numbers. Much the same is done in quantum theory. We require predictions of the values a_n and p_n. For this purpose, we need the following.

1. Introduction of mathematical entities such as an operator **A** which is associated with measurable physical properties of the system, and a state $|s\rangle$ which describes the physical condition of the system.
2. Postulates that relate the operators and states with the predicted values (a_n and p_n) of a measurement.
3. A body of mathematical tools to wield these abstract entities.

6.2 STATE VECTORS

Each single value (eigenvalue) a_n is associated with a single state (eigenstate) $|a_n\rangle$ that characterizes the state of the system after a measurement of **A** yields the eigenvalue a_n. Since an entire set of eigenvalues a_n are the possible results of experiment, this admits an entire set of associated eigenstates. If it is known beforehand that the system is in eigenstate $|a_4\rangle$, for instance, a measure of **A** furnishes the eigenvalue a_4. Any subsequent measure of **A** on this system will again yield the value a_4. Thus, once a system is in an eigenstate, it is not disturbed by the measurement but remains in that eigenstate. When, however, measurements are made on many identically prepared systems, each in a similar but unknown state, each measurement yields some eigenvalue. Given enough measurements, eventually all the eigenvalues a_n will have shown themselves, each with their own probability of occurrence p_n. It is as if (before any one measurement) the system were in a state composed of varying amounts of all the eigenstates, and a measurement caused the system to leap into one of the eigenstates. Since it is impossible to predict which eigenstate will be leapt into prior to the measurement the best we can say is that there is a probability p_n for ending up in the eigenstate $|a_n\rangle$.

We *assume* that once having specified the complete set of eigenstates, we can construct a general state of the system as some simple sum of all the eigenstates. The simplest combination is a linear one where the eigenstates enter into a sum each with a coefficient c_n as in

$$|s\rangle = \sum_n c_n|a_n\rangle \tag{6.1}$$

It is a provision of the theory that the coefficients c_n are generally complex rather than simply real numbers. This necessity for a linear superposition of eigenstates is really one of the basic *assumptions* of the theory, since it enables, through the set of coefficients and eigenstates, a simple and unique construction of any arbitrary state.

We do not usually know beforehand what exactly is the state of the system. It may be in any of the eigenstates or, as is more often the case, the system will be in a general state $|s\rangle$ which is a combination of some or all of the possible eigenstates. This is where the probability enters. We postulate that the absolute square of the coefficients c_n of each eigenstate is the probability

$$p_n = c_n^* c_n = |c_n|^2 \tag{6.2}$$

of finding the eigenvalue c_n in a single measurement. One problem of quantum theory, then, is to find these coefficients; with them, we can predict the probabilities p_n.

The entire set of coefficients c_n appear all at once in the arbitrary state

$$|s\rangle = c_0|a_0\rangle + c_1|a_1\rangle + \cdots + c_n|a_n\rangle + \cdots \tag{6.3}$$

It is expedient, therefore, to develop some mathematical tools which can isolate the components of this general state, and thereby expose some of the mathematical properties and relations among the coefficients and eigenstates. For this purpose an analogy between these abstract states and ordinary vectors is useful.

Assigning a direction along an axis to each of the eigenstates $|a_n\rangle$, we can make an analogy between state space and vector space as defined by unit vectors. In two dimensions, for example, any vector can be constructed

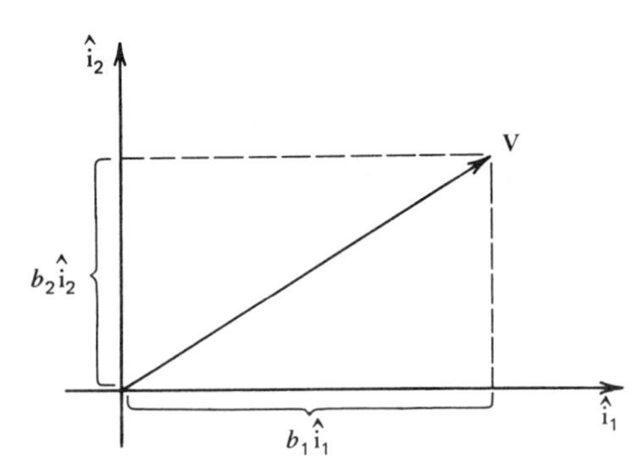

Figure 6.3. Composition of a two-dimensional vector.

by choosing the right coefficients of each unit vector $\hat{\mathbf{i}}_1$ and $\hat{\mathbf{i}}_2$. Thus, a vector **V** becomes (Fig. 6.3)

$$\begin{aligned}\mathbf{V} &= b_1\hat{\mathbf{i}}_1 + b_2\hat{\mathbf{i}}_2 \\ &= \sum_{n=1}^{2} b_n\hat{\mathbf{i}}_n \end{aligned} \tag{6.4}$$

This is similar to the superposition of eigenstates where, in the case of two-eigenstate "space" (Fig. 6.4), a general state vector, as we shall now call it, is given by

$$\begin{aligned}|s\rangle &= c_1|a_1\rangle + c_2|a_2\rangle \\ &= \sum_{n=1}^{2} c_n|a_n\rangle \end{aligned} \tag{6.5}$$

The coefficients c_n appear to be the magnitude of the projection of the state vector $|s\rangle$ on the eigenstate vector (eigenvector) direction—in keeping with the ideas concerning ordinary vectors. To get at the coefficients, which are the projections of ordinary vectors on their unit vectors, we use the dot product

$$b_n = \hat{\mathbf{i}}_n \cdot \mathbf{V} \tag{6.6}$$

Following this idea, we invent a "dot product" for state vectors having the sense of a projection of one state vector on another, which in the particular case of projecting the state vector $|s\rangle$ onto the eigenvector $|a_n\rangle$, gives the

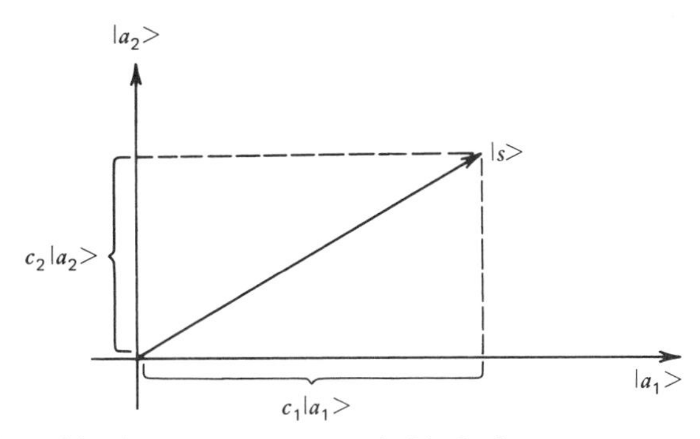

Figure 6.4. A state vector composed of its basic component states.

coefficient c_n. Instead of defining this product as

$$c_n = |a_n\rangle \cdot |s\rangle \tag{6.7}$$

we must, since these state vectors deal with complex numbers, define the projection as

$$c_n = |a_n\rangle^\dagger \cdot |s\rangle \tag{6.8}$$

where $|a_n\rangle^\dagger$, called the adjoint of $|a_n\rangle$, is related to $|a_n\rangle$ by the further definition

$$|a_n\rangle^\dagger \equiv \langle a_n| \tag{6.9}$$

and

$$|a_n\rangle^{\dagger\dagger} = \langle a_n|^\dagger = |a_n\rangle \tag{6.10}$$

This action of taking the adjoint is somewhat analogous to taking a complex conjugate, except that state vectors, while they deal with complex coefficients, are not exactly themselves complex numbers.

The state vector product can now be written, with the aid of the adjoint definitions, as

$$c_n = \langle a_n| \cdot |s\rangle \tag{6.11}$$

In keeping with standard notation, the dot and one of the two bars in the center is eliminated for economy, making the state vector product

$$c_n = \langle a_n|s\rangle \tag{6.12}$$

The projection, therefore, of one state vector on a second is the adjoint of the second times the first. We could, however, have equally well defined the product as

$$c_n^* = |s\rangle^\dagger|a_n\rangle = \langle s|a_n\rangle \tag{6.13}$$

which differs from the former product $\langle a_n|s\rangle$ in that one is the complex conjugate of the other, that is

$$c_n = \langle a_n|s\rangle = \langle s|a_n\rangle^* \tag{6.14}$$

It is interesting to note, that up until now the discussion about states of a system has been in terms of the state vectors $|a_n\rangle$ and $|s\rangle$ (called *kets*). The states could just as well have been discussed in terms of the adjoints $\langle a_n|$ and $\langle s|$ (called *bras*).[‡] As an example, consider that either

$$|s\rangle = \sum_n c_n |a_n\rangle \tag{6.15}$$

or the adjoint[§]

$$\langle s| = \left(\sum_n c_n |a_n\rangle\right)^\dagger = \sum_n c_n^* \langle a_n| \tag{6.16}$$

related by

$$\langle s|^\dagger = |s\rangle \tag{6.17}$$

both describe the same general state vector.

6.3 PROBABILITIES AND STATES

Returning to the postulate concerning the probability p_n, we can now couch its coefficients in terms of state vector products. Witness the expression

$$p_n = c_n^* c_n = \langle a_n|s\rangle^* \langle a_n|s\rangle \tag{6.18}$$

which as the product of a complex number and its complex conjugate is

$$p_n = |c_n|^2 = |\langle a_n|s\rangle|^2 \tag{6.19}$$

Knowing, thus, the magnitude of the state vector product, makes it possible to predict the probability for "getting to" the eigenstate $\langle a_n|$ for a system initially in the state $|s\rangle$. Part of the quantum description of a system, therefore, depends upon a prudent choice for the values

$$c_n = \langle a_n|s\rangle \tag{6.20}$$

In actual practice, once the eigenstates are known for a particular system, a

[‡]The names *bra* and *ket*, invented by Dirac, come from the association with the word bracket (bra–ket).

[§]The adjoint of a complex number is its complex conjugate: $c_n^\dagger = c_n^*$.

convenient choice of a state $|s\rangle$ is made, and the coefficients $c_n = \langle a_n|s\rangle$ are then calculated. If the square of these coefficients gives a probability, $p_n = |c_n|^2$, which is in agreement with the experimental results, we have indeed made not only a convenient but also an excellent choice for the general state vector $|s\rangle$.

Considering that a system may be in any arbitrary state, we may quite possibly start with a system already in an eigenstate, in which case the state vector $|s\rangle$ is specifically the eigenstate, say, $|a_m\rangle$. The probability of finding the eigenvalue a_n for this system in the specific eigenstate $|a_m\rangle$ is *zero*. Only the value a_m is a possible result of a measurement on a system in the state $|a_m\rangle$. On the other hand, the probability of finding the value a_m for the eigenstate $|a_m\rangle$ is *unity*, since a measure of the property **A** will always produce the eigenvalue a_m for a system in the eigenstate $|a_m\rangle$. Another way of saying this is

$$p_n = |\langle a_n|a_m\rangle|^2 = 0 \tag{6.21}$$

when $n \neq m$, and

$$p_n = |\langle a_n|a_m\rangle|^2 = 1 \tag{6.22}$$

when $n = m$, from which we *deduce* for the former

$$\langle a_n|a_m\rangle = 0 \qquad n \neq m \tag{6.23}$$

and *assume*‡ for the latter

$$\langle a_n|a_m\rangle = 1 \qquad n = m \tag{6.24}$$

Once again, reflecting upon the vector analogy that the product of two perpendicular vectors is zero, we use condition (6.23) to define the idea of orthogonality for eigenvectors. Two states are orthogonal when their state vector product is zero. Choosing $\langle a_n|a_n\rangle = 1$ is a way of preparing unit eigenstates (in the sense of unit vectors) whose lengths are fixed. With ordinary vectors their absolute length is of physical importance and is therefore preserved. It is the "direction" of state vectors, on the other hand, that is of physical significance, allowing their magnitudes by *definition* to be normalized, that is,

$$\langle a|a\rangle = 1 \tag{6.25}$$

‡In this assumption there is the twofold implication that (1) the magnitude of the complex number, $\langle a_m|a_m\rangle = |\mathbf{r}|e^{i\theta}$, can be set equal to one; (2) the phase can be chosen as zero.

6.4 OPERATORS AND OPERATIONS

Without naming it, we have made repeated reference to a physical property **A**, the measure of which leads to one of the eigenvalues. It is an *assumption* of the theory that with each measurable physical variable of a system, there exists or is assigned a mathematical entity called an operator that is associated with that physical, or as it is classically referred to, *dynamical variable*. Operators, like states, are another kind of mathematical gadgetry whose presence is justified by the final results. Given the rules of operation for operators, which are only remotely if at all connected with physical ideas, predictions from quantum theory in which the operators enter are in agreement with experimental observations. Symbolized simply by the notation **A**, the operator, unlike ordinary numbers or functions, is without mathematical meaning until it operates on a state.

As with the analogy of ordinary vectors to state vectors, we can conjure up an analogy between ordinary operations and these quantum operators. A familiar mathematical operation, for example, is performed in multiplying a function $f(x)$ by x. Thus

$$x f(x) = g(x) \tag{6.26}$$

can be looked upon as the operator x acting on the function $f(x)$, transforming it into $g(x)$. Another example which better demonstrates the abstractness of an operator is seen in taking the derivative

$$\frac{\partial}{\partial x} f(x) = g(x) \tag{6.27}$$

Here, the operator

$$\mathbf{A} = \frac{\partial}{\partial x} \tag{6.28}$$

has meaning only in terms of its operation on a function.

$$\mathbf{A} f(x) = g(x) \tag{6.29}$$

An operator in quantum theory acts most generally on a state to produce another state, namely

$$\mathbf{A}|s\rangle = |s'\rangle \tag{6.30}$$

Similarly, the adjoint of this relation is defined as

$$(\mathbf{A}|s\rangle)^\dagger = \langle s|\mathbf{A}^\dagger = \langle s'| \tag{6.31}$$

with the stipulation that $\mathbf{A}^{\dagger\dagger} = \mathbf{A}$. Also, we adopt the convention that operators are placed to the right of the bra or the left of the ket on which they act.

If we examine the product between states of $|s'\rangle$ and $|r\rangle$ we have either

$$\langle r|s'\rangle = \langle r|(\mathbf{A}|s\rangle) \quad \text{or} \quad \langle s'|r\rangle = (\langle s|\mathbf{A}^\dagger)|r\rangle \tag{6.32}$$

Since one product is the complex conjugate of the other, namely,

$$\langle r|s'\rangle^* = \langle s'|r\rangle \tag{6.33}$$

it can be shown that

$$\langle r|\mathbf{A}|s\rangle = \langle s|\mathbf{A}^\dagger|r\rangle^* \tag{6.34}$$

where now the $\mathbf{A}$ and $\mathbf{A}^\dagger$ act forward or backward on either state. Hence, there is no need for the parentheses. The product of $|s'\rangle$ and $|r\rangle$ may be written as

$$\langle r|s'\rangle = \langle r|\mathbf{A}|s\rangle \tag{6.35}$$

where the operator $\mathbf{A}$ may act either way.

We have referred throughout the chapter to an operator $\mathbf{A}$, which we will call an *observable operator*, as being associated with a measurable dynamical variable. There is, however, a fundamental requirement for an operator to be observable. It must be Hermitian, that is, the operator must be its own adjoint.

$$\mathbf{A} = \mathbf{A}^\dagger \tag{6.36}$$

The reason for this, as will be seen shortly, is that only Hermitian operators lead to real eigenvalues, as they must if they are to concur with real experimental results. Thus, Hermitian operators occupy a favored position in the theory.

6.5 POSTULATES AND EXPECTATIONS

We are now in a position to examine other postulates of quantum theory with whose help the eigenvalues and probabilities can be calculated for particular physical systems. In classical physics, prediction of the measured value of a variable and the mathematical entity representing that variable are one and the same thing, namely, ordinary numbers. In quantum theory, predictions of the measured value (eigenvalue) and the mathematical entity assigned to its dynamical variable (Hermitian operator) must be related by an additional *postulate*.

If an Hermitian operator acts on a state of a system to produce another state which differs from the original by at most a numerical factor (real number), then that numerical factor is an eigenvalue and the state is an associated eigenstate. In mathematical language this postulate resolves that

$$\mathbf{A}|a_n\rangle = a_n|a_n\rangle \tag{6.37}$$

where we use the practice of labeling the eigenvector with the eigenvalue. Equation (6.37), called the eigenvalue equation, is of central importance to the theory since its solution determines the eigenvalues a_n and the eigenstates $|a_n\rangle$.

It can be shown that the eigenvalues a_n of the Hermitian operator **A** are real numbers, as need be, since they predict real measurable values. Furthermore, it is possible to demonstrate that the eigenvector solutions are all orthogonal and, upon normalizing, can be expressed as

$$\langle a_m|a_n\rangle = \begin{cases} 1 & \text{for } n = m \\ 0 & \quad n \neq m \end{cases} \tag{6.38}$$

Putting these features together with the general properties of statevectors and operators established thus far, we can write the mathematical statement that predicts the measured value of an operator. Let us form the product of $\langle a_n|$ with the eigenvalue equation (6.38). This gives

$$\langle a_n|\mathbf{A}|a_n\rangle = \langle a_n|a_n|a_n\rangle \tag{6.39}$$

where the right-hand side of the equation can be simplified (since a_n is just an ordinary number) to

$$\langle a_n|\mathbf{A}|a_n\rangle = a_n\langle a_n|a_n\rangle \tag{6.40}$$

Using the orthogonality property of the eigenvectors, we can write

$$\langle\mathbf{A}\rangle = \langle a_n|\mathbf{A}|a_n\rangle = a_n \tag{6.41}$$

with the understanding that $\langle\mathbf{A}\rangle$ means the measured value of the operator **A**, which for this system in the eigenstate $|a_n\rangle$, has the value

$$\langle\mathbf{A}\rangle = a_n \tag{6.42}$$

While the above definition of the value of an operator suffices for a system in an eigenstate, it must be revised somewhat to be meaningful for systems in a more general state.

Let us suppose that, having solved the eigenvalue equation for the eigenstates $|a_n\rangle$, the system is in a general state

$$|s\rangle = \sum_n c_n |a_n\rangle \tag{6.43}$$

which, as previously defined, is some combination of the eigenstates. We define the expectation value of the operator **A** for this state as

$$\langle \mathbf{A} \rangle = \langle s|\mathbf{A}|s\rangle = \Big(\sum_m c_m^* \langle a_m|\Big)\mathbf{A}\Big(\sum_n c_n |a_n\rangle\Big) \tag{6.44}$$

which, with the aid of the eigenvalue equation (6.37) and the orthogonality relationships (6.38), becomes

$$\langle \mathbf{A} \rangle = \sum_n p_n a_n \tag{6.45}$$

where $p_n = c_n^* c_n$ is the probability as previously discussed.

For a system in an eigenstate, the value of the operator **A** has the meaning that a measure of $\langle \mathbf{A} \rangle$, for the eigenstate $|a_n\rangle$, yields the value

$$\langle \mathbf{A} \rangle = a_n \tag{6.46}$$

In other words, the eigenvalue a_n is always the result of a measurement of **A** for a system in an eigenstate $\langle a_n|$. Another measurement on the same system will again yield a particular a_n. This is not so, however, for a system in a general state $|s\rangle$. A measure of **A** on this system will yield some one of the possible eigenvalues a_n, say a_m, but it is impossible to know in advance which one. The most we can say is that if many systems, all prepared in the same state $|s\rangle$, were each measured for **A**, the probable occurrence of the eigenvalue a_n would be p_n. Since all eigenvalues are possible results of the measurements, each with a probability p_n, the value of **A** for an average of all the measurements, then, is

$$\langle \mathbf{A} \rangle = \sum_n p_n a_n \tag{6.47}$$

But this is just what we derived mathematically from

$$\langle \mathbf{A} \rangle = \langle s|\mathbf{A}|s\rangle \tag{6.48}$$

Therefore, for a system in a state $|s\rangle$, the measured expectation value of the operator $\langle \mathbf{A} \rangle$ has the meaning of the average of many measurements on similar systems. Ideally, of course, we should make measurements on an infinite number of identical systems, the collection of which is called an *ensemble*. This quantum average, unlike classical statistical averages, is an integral part of quantum theory, and as such, provides the best possible description for the behavior of a physical system. In a particular problem, therefore, it is the expectation values that we would like to calculate as part of the predicted description of the system. To do so, it is necessary first to solve the eigenvalue equation.

6.6 QUANTIZATION

We have dealt rather generally with eigenvalues, states, and operators, never really tying them to an actual physical property. Attention will now be focused specifically on momentum, position, and energy. For example, if we are interested in the eigenvalues and eigenstates of the energy operator **H** (which is called the Hamiltonian), it is necessary to solve the eigenvalue equation

$$\mathbf{H}|E_n\rangle = E_n|E_n\rangle \tag{6.49}$$

where E_n and $|E_n\rangle$ are, respectively, the energy eigenvalues and energy eigenstates of the Hamiltonian operator **H**.

As might be anticipated (if it is possible to anticipate anything with all the surprises of quantum theory), the Hamiltonian operator for a point mass is expressible as a function of the momentum and position operators, $\mathbf{H}(\mathbf{p}, \mathbf{q})$. We proceed to a postulate concerning momentum and position operators that makes possible the solution of the energy eigenvalue problem. Although most abstract, this postulate is also most profound, since it introduces Planck's constant, leads to Heisenberg's uncertainty principle,[‡] and exposes both the *correspondence* and profound difference between quantum theory and classical theory.

Mathematically, the postulate can be expressed very simply. The momentum operator **p** and position operator **q** satisfy the relation

$$\mathbf{qp} - \mathbf{pq} = i\hbar \tag{6.50}$$

This statement, in its simplicity, is matched only by its abstractness but it is

[‡]The uncertainty principle is inherent in all quantum results. Since it is not used explicitly in what follows, it is relegated to separate discussion in subsequent chapters.

the thread from which all quantum theory hangs. It is this requirement (6.50) that *allows for the quantum effects* through the slender difference in value between $\hbar$ and zero.[‡] For if $\hbar$ were to approach zero, the operator relation would become

$$\mathbf{qp} - \mathbf{pq} \to 0 \tag{6.51}$$

which in the limit would be

$$\mathbf{qp} = \mathbf{pq} \tag{6.52}$$

This equality is the commutative property, that is, the order of multiplication is immaterial, and **p** and **q** are said to commute with one another. While this is so in classical physics for momentum and position, in quantum theory **p** and **q** are noncommuting operators saved by the slim margin of the constant, $\hbar$.

To illustrate the importance of the smallness of $\hbar$, let us consider a basic unit of the theory, a quantum of energy

$$E_0 = \hbar\omega \tag{6.53}$$

This energy, except for extremely high frequencies ω, is exceedingly tiny compared to the energies encountered in classical macroscopic bodies. A system having a large energy E (large compared to $\hbar\omega$) is practically unaffected by quantum size effects having energies of the order of $\hbar\omega$; it is, therefore, describable by classical physics. Another way of looking at it is as follows: From high up in the energy scale, where $E \gg \hbar\omega$, the quantity $\hbar\omega$ and, therefore, $\hbar$ appear to be numbers very close to zero. On the other hand, systems having small energies (on the order of $\hbar\omega$, as encountered with atoms) behave in their peculiar quantum way and must be described by quantum, not classical expectations.

It is a welcome correspondence (the correspondence principle) that quantum predictions for large energies must approach the classical description. This provides a check on quantum theory (at least in that limit) and some possible guidance about the structure of quantum operators. The actual construction of a quantum operator is at best a calculated guess guided by its classical counterpart (if there is one) and justified in the end only by its ability to lead to predictions that agree with experimental results.

[‡] $2\pi\hbar = h = 6.625 \times 10^{-34}$ J s.

6.7 HAMILTONIAN OPERATOR AND HARMONIC OSCILLATOR

As an important case in point, let us consider the classical energy of a point mass m. The classical Hamiltonian H for this mass having a momentum p and a position q in some external force field is given by

$$H = \frac{p^2}{2m} + V(q) \tag{6.54}$$

where $V(q)$ is the potential energy of the particle in the field at the position q. In quantum theory this Hamiltonian operator takes the form of its classical counterpart. The operator is assumed to be

$$\mathbf{H} = \frac{\mathbf{p}^2}{2m} + V(\mathbf{q}) \tag{6.55}$$

where $\mathbf{p}$ and $\mathbf{q}$ are momentum and position operators, respectively. Using it in the energy eigenvalue equation

$$\left[\frac{\mathbf{p}^2}{2m} + V(\mathbf{q})\right]|E_n\rangle = E_n|E_n\rangle \tag{6.56}$$

together with the condition, $\mathbf{qp} - \mathbf{pq} = i\hbar$, it becomes possible to solve for the energy eigenvalues E_n and the accompanying energy eigenstates $|E_n\rangle$. Of course, the potential energy operator $V(\mathbf{q})$ must be specified for the problem. Here again the calculated guess is that it has a form similar to its classical analogue.

In the specific example of simple harmonic motion, where a unit mass ($m = 1$) is on the end of a spring having a spring constant K, the potential energy is

$$V(q) = \tfrac{1}{2}Kq^2 \tag{6.57}$$

The Hamiltonian operator for this system, therefore, is assumed to be

$$\mathbf{H} = \tfrac{1}{2}(\mathbf{p}^2 + \omega^2\mathbf{q}^2) \tag{6.58}$$

where we used $K = \omega^2 m$ for the oscillator (ω is the frequency) and $m = 1$. With this operator, we can attend to the solution of the eigenvalue equation

$$\tfrac{1}{2}(\mathbf{p}^2 + \omega^2\mathbf{q}^2)|E_n\rangle = E_n|E_n\rangle \tag{6.59}$$

which specifies the eigenvalues and eigenvectors of the simple harmonic oscillator. When attained, these quantities become useful for the following:

1. Construction of an arbitrary state where the coefficients c_n are given by

$$c_n = \langle E_n | s \rangle \tag{6.60}$$

2. Calculation of the expectation energy for the arbitrary state $|s\rangle$ as defined by

$$\langle \mathbf{H} \rangle = \langle s | \mathbf{H} | s \rangle = \sum_n p_n E_n \tag{6.61}$$

It is the states $|s\rangle$ and their average energy $\langle \mathbf{H} \rangle$ with which we will be concerned in detail. But in order to arrive at the constructions and calculations, we must know the energy eigenvalues and eigenstates of the harmonic oscillator. The next chapter proceeds to this task.

REFERENCES

Bohm, D. *Quantum Theory*, Prentice-Hall, New York, 1951.

Dirac, P. A. M. *The Principles of Quantum Mechanics*, Oxford University Press, Oxford, England, 1955.

Louisell, W. H. *Radiation and Noise in Quantum Electronics*, McGraw-Hill, New York, 1964.

Messiah, A. *Quantum Mechanics*, North-Holland, Amsterdam, 1958.

Powell, J. L. and B. Crasemann. *Quantum Mechanics*, Addison–Wesley, Reading, Mass., 1961.

Schiff, L. I. *Quantum Mechanics*, McGraw–Hill, New York, 1955.

PROBLEMS

6.1. Show that the adjoint of a complex number c_n is its complex conjugate.

6.2. Prove that $\langle r | \mathbf{A} | s \rangle = \langle s | \mathbf{A}^\dagger | r \rangle^*$.

6.3. Show that the eigenvalues of a Hermitian operator are real.

6.4. Using the commutator definition $[\mathbf{q}, \mathbf{p}] = \mathbf{qp} - \mathbf{pq}$ shows the results $[\mathbf{q}^n, \mathbf{p}] = i\hbar n \mathbf{q}^{n-1}$ and $[\mathbf{q}, \mathbf{p}^n] = i\hbar n \mathbf{p}^{n-1}$.

6.5. Writing an arbitrary state of a system in terms of its eigenstates as $|\psi\rangle = \sum_n c_n |n\rangle$, if the eigenkets $|n\rangle$ are all orthogonal to one another, show that $\sum_n |n\rangle\langle n| = 1$.

6.6. Verify that the following is true:

$$\left[x, \frac{d}{dx}\right] f(x) = -f(x)$$

6.7. If $\mathbf{AA}^{-1} = 1$ defines the relation of an operator and its inverse prove that $(\mathbf{ABC})^{-1} = \mathbf{A}^{-1}\mathbf{B}^{-1}\mathbf{C}^{-1}$.

6.8. For a free particle show that $[\mathbf{p}, \mathbf{H}] = 0$.

6.9. Using $\psi(q) = \langle q|\psi\rangle$, show that $\mathbf{H}|\psi\rangle = E|\psi\rangle$ becomes the time-independent Schrödinger equation.

6.10. Show that two state vectors $|a\rangle$ and $|b\rangle$ satisfy the relation $|\langle a|b\rangle|^2 \leqslant \langle a|a\rangle\langle b|b\rangle$ and that the equality sign is true only if $|a\rangle = c|b\rangle$, where c is constant.

7

The Quanta and Coherent States of an Oscillator

A quantum mechanical harmonic oscillator is distinguished by its Hamiltonian operator

$$\mathbf{H} = \tfrac{1}{2}(\mathbf{p}^2 + \omega^2\mathbf{q}^2) \tag{7.1a}$$

where $\mathbf{H}$, $\mathbf{p}$, and $\mathbf{q}$ are the momentum and position operators for an oscillator of unit mass. It has energy eigenvalues E_n and associated energy eigenstates $|E_n\rangle$ specified by the eigenvalue equation

$$\tfrac{1}{2}(\mathbf{p}^2 + \omega^2\mathbf{q}^2)|E_n\rangle = E_n|E_n\rangle \tag{7.1b}$$

A solution of this equation for the eigenvalues and eigenstates can be achieved algebraically with the aid of the commutation postulate

$$\mathbf{pq} - \mathbf{qp} = -i\hbar \tag{7.2}$$

and some of the basic operator–state properties. To facilitate the procedure, we will find it convenient to define a new operator and its adjoint, as combinations of the momentum and position operators, in terms of which the eigenvalue equation takes on a simple algebraic formulation.

7.1 FORMULATING THE EIGENVALUE PROBLEM

The operators $\mathbf{a}^\dagger$ and $\mathbf{a}$, named *raising and lowering operators* for reasons which will be obvious later, are defined by

$$\mathbf{a}^\dagger = (2\hbar\omega)^{-1/2}(\omega\mathbf{q} - i\mathbf{p}) \quad \text{and} \quad \mathbf{a} = (2\hbar\omega)^{-1/2}(\omega\mathbf{q} + i\mathbf{p}) \tag{7.3}$$

These relations are already familiar in that they are mathematically similar to those of the complex time-dependent amplitudes introduced in the electromagnetic field formulation (Section 5.2). The difference here is that $\mathbf{a}^\dagger$ and $\mathbf{a}$ are noncommuting operators defined by combinations of position and momentum operators, rather than complex time-dependent functions defined in terms of time-dependent field amplitudes.

Using the commutation property of $\mathbf{p}$ and $\mathbf{q}$, the raising and lowering operators obey the commutation relation

$$\mathbf{a}\mathbf{a}^\dagger - \mathbf{a}^\dagger\mathbf{a} = 1 \tag{7.4}$$

This enables the oscillator Hamiltonian to be rewritten as

$$\mathbf{H} = \tfrac{1}{2}\left(\mathbf{p}^2 + \omega^2\mathbf{q}^2\right) = \hbar\omega\left(\mathbf{a}^\dagger\mathbf{a} + \tfrac{1}{2}\right) \tag{7.5}$$

for which the eigenvalue equation becomes

$$\hbar\omega\left(\mathbf{a}^\dagger\mathbf{a} + \tfrac{1}{2}\right)|E_n\rangle = E_n|E_n\rangle \tag{7.6}$$

or more conveniently

$$\mathbf{a}^\dagger\mathbf{a}|E_n\rangle = \left(\frac{E_n}{\hbar\omega} - \frac{1}{2}\right)|E_n\rangle \tag{7.7}$$

By introducing the number operator

$$\mathbf{N} = \mathbf{a}^\dagger\mathbf{a} \tag{7.8a}$$

and the eigenvalue number[‡]

$$n = \frac{E_n}{\hbar\omega} - \frac{1}{2} \tag{7.8b}$$

and by relabeling the eigenstate $|E_n\rangle = |n\rangle$ the eigenvalue equation becomes simply

$$\mathbf{N}|n\rangle = n|n\rangle \tag{7.9}$$

We are still dealing with the harmonic oscillator except that instead of the Hamiltonian operator

$$\mathbf{H} = \hbar\omega\left(\mathbf{a}^\dagger\mathbf{a} + \tfrac{1}{2}\right) \tag{7.10}$$

[‡]As introduced here, n is not required to be an integer although it will represent integers later.

we are using **N**, the number operator, whose eigenvalues are related to the energy eigenvalues by

$$E_n = \hbar\omega\left(n + \tfrac{1}{2}\right) \tag{7.11}$$

There are restrictions on the value of n, as will be shown. An operator acting on a state produces another state

$$\mathbf{a}|n\rangle = |m\rangle \tag{7.12}$$

or in terms of the adjoint

$$\langle n|\mathbf{a}^{\dagger} = \langle m| \tag{7.13}$$

The product of the state with itself is

$$\langle n|\mathbf{a}^{\dagger}\mathbf{a}|n\rangle = \langle m|m\rangle \tag{7.14}$$

which must be greater than or equal to zero.‡ In the case where $\mathbf{a}^{\dagger}$, $\mathbf{a}$, and $|n\rangle$ are the operators and eigenstates of the oscillator, then

$$\mathbf{a}^{\dagger}\mathbf{a}|n\rangle = n|n\rangle \tag{7.15}$$

The product of the states Eq. (7.14) becomes

$$\langle n|\mathbf{a}^{\dagger}\mathbf{a}|n\rangle = \langle n|\mathbf{N}|n\rangle = n\langle n|n\rangle = n \geqslant 0 \tag{7.16}$$

where the normalization constant for $\langle n|n\rangle$ has been set equal to one. In other words the eigenvalue n for the harmonic oscillator cannot be less than zero. The next step is to solve the simplified eigenvalue equation.

7.2 EIGEN-NUMBERS THROUGH ALGEBRA

Now here comes the surprising algebraic simplicity! Consider what happens when we form the following:

1. Operate with **a** on both sides of the eigenvalue equation for **N** to give

$$\mathbf{a}\mathbf{a}^{\dagger}\mathbf{a}|n\rangle = n\mathbf{a}|n\rangle \tag{7.17}$$

‡The expression $\langle m|m\rangle$ can be normalized to any finite positive value not zero.

2. The commutation property of the raising and lowering operators can be rewritten as

$$\mathbf{aa}^\dagger = (1 + \mathbf{aa}) \tag{7.18}$$

which allows the eigenvalue relation to become

$$(1 + \mathbf{a}^\dagger\mathbf{a})\mathbf{a}|n\rangle = n\mathbf{a}|n\rangle \tag{7.19}$$

3. Carry out an algebraic rearrangement separating the operators and states to read

$$\mathbf{a}^\dagger\mathbf{a}[\mathbf{a}|n\rangle] = (n - 1)[\mathbf{a}|n\rangle] \tag{7.20}$$

This equation is still the eigenvalue equation but now more explicitly for the eigenvalue $(n - 1)$ of a new eigenstate $\mathbf{a}|n\rangle$. That is, the quantity $\mathbf{a}|n\rangle$ is just another eigenstate of the oscillator. In keeping with our labeling practice we may write

$$\mathbf{a}|n\rangle = b_n|n - 1\rangle \tag{7.21}$$

where b_n is a complex number which might be needed to normalize the state $|n - 1\rangle$.

This result provides the clue to solving for the eigenvalues. Since the operator $\mathbf{a}$ acting on an eigenstate lowers its index by exactly one, an entire set of eigenstates, and therefore eigenvalues, can be generated. To illustrate a few, consider the following:

$$\begin{aligned}
\mathbf{a}|n\rangle &= b_n|n - 1\rangle \\
\mathbf{a}|n - 1\rangle &= b_{n-1}|n - 2\rangle \\
\vdots \quad & \quad \vdots \\
\mathbf{a}|n - m\rangle &= b_{n-m}|n - (m + 1)\rangle
\end{aligned} \tag{7.22}$$

This lowering process has to stop somewhere since energy eigenvalues cannot become negative. There must be an eigenstate for which $\mathbf{a}|n_{\text{lowest}}\rangle$ does not generate another lower one. This will be true if

$$\mathbf{a}|n_{\text{lowest}}\rangle = 0 \tag{7.23}$$

We have an earlier result

$$\langle n|\mathbf{a}^\dagger\mathbf{a}|n\rangle = n \tag{7.24}$$

which requires that when $n = 0$,

$$\mathbf{a}|0\rangle = 0 \tag{7.25}$$

Combining these two results, we have that $|n_{\text{lowest}}\rangle$ must be the eigenstate $|0\rangle$. This eigenstate is the ground state of the oscillator and has an associated eigenvalue $n = 0$. The other eigenvalues must be all the integers above zero (Fig. 7.1). As might be intuited, it can be shown that the adjoint operator performs the raising operation

$$\mathbf{a}^{\dagger}|n - 1\rangle = b_n^*|n\rangle \tag{7.26}$$

from which it follows that there is no upper bound on the eigenvalues.

From the definition of n in terms of the energy eigenvalues E_n of the harmonic oscillator, we have finally the complete set of values

$$E_n = \hbar\omega\left(n + \tfrac{1}{2}\right) \tag{7.27}$$

where $n = 0, 1, 2, \ldots, m, \ldots$. This quantum result sets the ground state energy of the oscillator at

$$E_0 = \tfrac{1}{2}\hbar\omega \tag{7.28}$$

with all succeeding allowable energies being equally spaced by the quantity $\hbar\omega$ up from the ground state energy (Fig. 7.2). Energies corresponding to noninteger numbers are simply forbidden, producing discontinuity in the

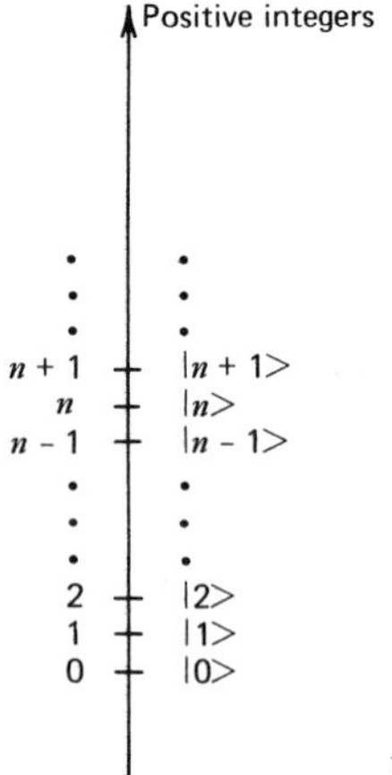

Figure 7.1. Eigenvalues of the oscillator and their associated states.

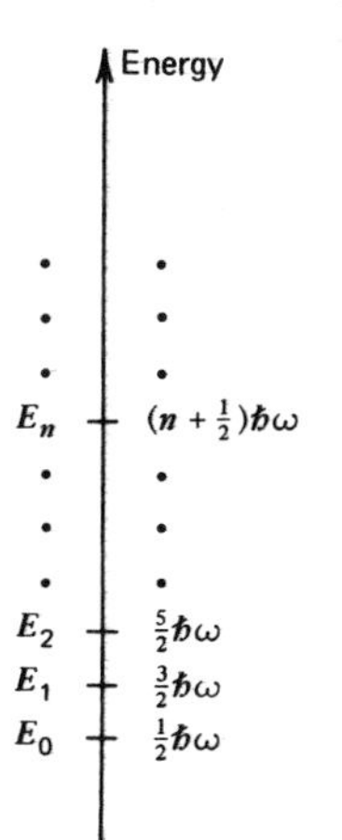

Figure 7.2. Scale of quantum energy levels for a harmonic oscillator.

allowed energy values for the oscillator. This result has no parallel with classical predictions but is uniquely related to the probabilistic approach and uncertainty built into the fabric of the quantum formulation.

7.3 RAISING AND LOWERING OPERATORS ON NORMALIZED STATES

While the number eigenstates of the harmonic oscillator must be all orthogonal, that is,

$$\langle m|n\rangle = 0 \qquad \text{for all } m \neq n \tag{7.29}$$

we have not yet taken the trouble to normalize them. We would like to have

$$\langle n|n\rangle = 1 \qquad \text{for all } n \tag{7.30}$$

This is done most easily by using the lowering operation

$$\mathbf{a}|n\rangle = b_n|n-1\rangle \tag{7.31}$$

and its adjoint statement

$$\langle n|\mathbf{a}^\dagger = b_n^*\langle n-1| \tag{7.32}$$

The product of the two gives

$$\langle n|\mathbf{a}^\dagger\mathbf{a}|n\rangle = b_n^* b_n\langle n-1|n-1\rangle = n \tag{7.33}$$

In order that the eigenstates $|n - 1\rangle$ be normalized, we require

$$b_n^* b_n = n \tag{7.34a}$$

Since the phase of the complex number b_n is arbitrary, it may be set equal to zero, making

$$b_n^* = b_n = \sqrt{n} \tag{7.34b}$$

Returning to the lowering and raising operations, we have finally

$$\mathbf{a}|n\rangle = \sqrt{n}\,|n - 1\rangle \tag{7.35}$$

and

$$\mathbf{a}^\dagger|n - 1\rangle = \sqrt{n}\,|n\rangle$$

Since each allowed energy differs from the next by a quantum of energy $\hbar\omega$, it is tempting to conceive of each eigenvalue E_n as composed of n quanta of energy plus the ground-state energy. While this interpretation is of small consequence here, it will prove to be important later when the mathematics of the oscillator is applied to the electromagnetic field.

The raising operator increases by one the number of the state on which it acts. This operation can be thought of as adding a quantum of energy $\hbar\omega$ to the state $|n - 1\rangle$, raising it to the state $|n\rangle$, as shown in Fig. 7.3. The opposite effect is produced by the lowering operator; a taking away of a quantum of energy lowers the state by one (Fig. 7.4).

The operators differ dramatically in their action on the ground state. In the ground state, there are no quanta left, hence, the lowering operator must yield

$$\mathbf{a}|0\rangle = 0 \tag{7.36}$$

Conversely, it is possible to add a quantum of energy to the ground state through the raising operation

$$\mathbf{a}^\dagger|0\rangle = \sqrt{1}\,|1\rangle \tag{7.37}$$

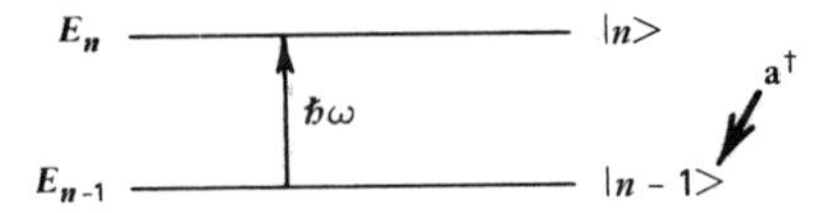

Figure 7.3. Action of the raising operator.

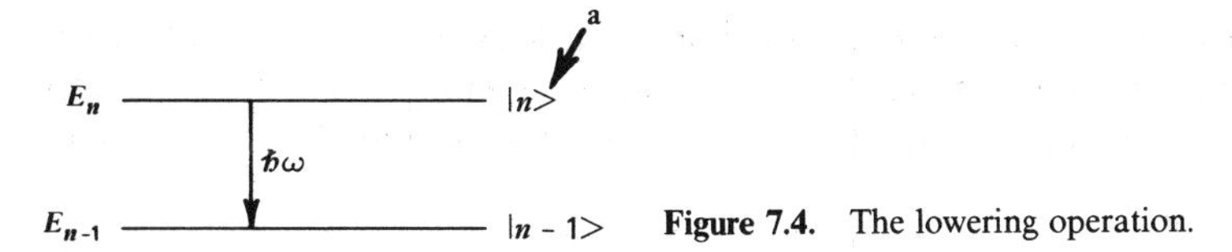

Figure 7.4. The lowering operation.

By repeated application of $\mathbf{a}^\dagger$ on the ground state, namely,

$$\mathbf{a}^{\dagger 2}|0\rangle = \sqrt{1}\,\mathbf{a}^\dagger|1\rangle = \sqrt{1}\sqrt{2}\,|2\rangle = \sqrt{2!}\,|2\rangle$$

$$\vdots$$

$$\mathbf{a}^{\dagger n}|0\rangle = \sqrt{1}\sqrt{2}\cdots\sqrt{n}\,|n\rangle = \sqrt{n!}\,|n\rangle \tag{7.38}$$

we obtain any state $|n\rangle$ expressed in terms of the ground state and raising operator, to be

$$|n\rangle = \frac{\mathbf{a}^{\dagger n}}{\sqrt{n!}}|0\rangle \tag{7.39}$$

This last expression tidily sums up the eigenstates of the harmonic oscillator, showing each to be but a raised version of the ground state. We now have a more general problem to consider; the construction of an arbitrary state of the oscillator and the calculation of the expectation value for the energy of that state. In other words, we wish to find the most likely quantum mechanical state of an ordinary oscillator.

7.4 THE MOST NATURAL STATE OF AN OSCILLATOR

Any arbitrary state is some combination of all the eigenstates, namely

$$|s\rangle = \sum_n c_n|n\rangle \tag{7.40}$$

The coefficients c_n are related to the probability of finding the eigenvalue n in a single measurement of the energy, that probability being expressed as

$$p_n = c_n^* c_n \tag{7.41}$$

The problem then becomes one of finding the probability p_n for our most likely ordinary oscillator.

For this purpose we take an oscillator and kick it, push it, or otherwise get it going in some state. A measurement of energy for this oscillator (assuming we could measure it accurately) must yield one of the eigenvalues

$$E_n = \left(n + \tfrac{1}{2}\right)\hbar\omega \tag{7.42}$$

While it is impossible to know before the measurement exactly which eigenvalue will occur, after a large number of measurements on identically prepared oscillators we will know the probability of occurrence p_n for each possible eigenvalue. Without prior knowledge of what the p_n actually are, the array of choices for the coefficients c_n is broad. To illustrate two extreme possibilities, consider

1. All $c_n = 0$, except one, say, c_m, which makes the state $|s\rangle$ reduce to the eigenstate $|m\rangle$. For an oscillator in this state, measurements of energy yield only the eigenvalue m all the time with a probability

$$p_m = c_m^* c_m = 1 \tag{7.43}$$

while all other probabilities p_n would be zero.

2. All coefficients c_n are equal, that is

$$c_1 = c_2 = \cdots = c_n = \cdots \tag{7.44}$$

in which case they form a state such that any eigenvalue at any time is *just as likely* to be found as any other since all probabilities

$$p_n = |c_1|^2 = |c_2|^2 = \cdots = |c_n|^2 = \cdots \tag{7.45}$$

are equal.

The former choice of coefficients describes an oscillator known to be in a definite eigenstate with a definite energy. The latter choice describes a state which expresses a complete lack of knowledge about the eigenstate in which the system will be found.

If our "kicked or otherwise pushed" oscillator always fell into one of these states, there would be no need to search further. This, however, is not the case. Take an actual classical oscillator; bang it into classical motion so that it oscillates sinusoidally with a single frequency ω and an average amplitude $\langle A\rangle$; it will most likely go into a state that lies somewhere between the two extremes.

The average energy of this classical oscillator

$$\langle E\rangle = \tfrac{1}{2}K\langle A\rangle^2 \tag{7.46a}$$

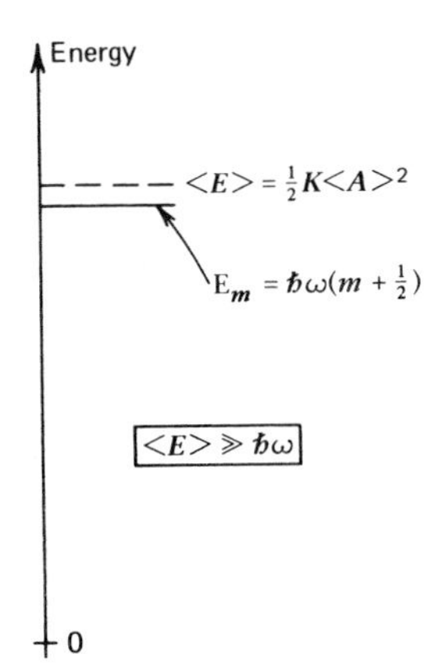

Figure 7.5. The average energy of a classical oscillator and one of its quantum eigenvalues.

is high on the energy scale where the eigenvalues are very large (Fig. 7.5). A measurement of its energy yields one of the eigenvalues, say,

$$E_m = \hbar\omega\left(m + \tfrac{1}{2}\right) \tag{7.46b}$$

If the same oscillator is prepared exactly as before, another measurement of its energy will yield, say, $E_{m'}$, which will probably again be close to the average energy but can be distinguished from both E_m and $\langle E \rangle$ (Fig. 7.6).

Repeating this procedure many times for identically prepared oscillators, we would observe from measurements of the energies, a group of eigenvalues, most of which would be very close to the average energy but some of which would have values further away. If we plot each single energy value against the number of observations of each eigenvalue, we obtain a picture of the distribution of the eigenvalues in terms of their frequency of occurrence (Fig. 7.7).

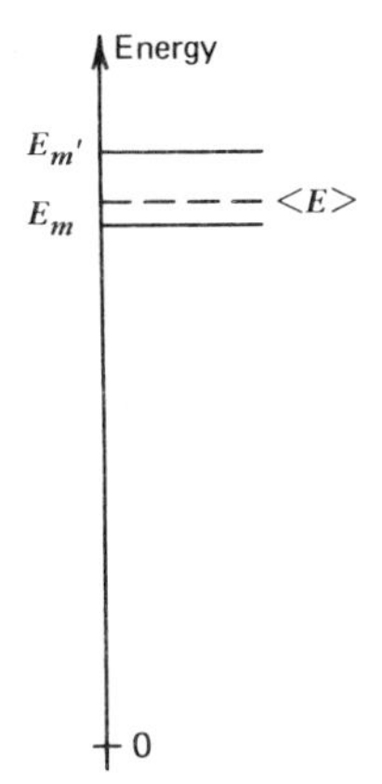

Figure 7.6. Another possible eigenvalue $E_{m'}$.

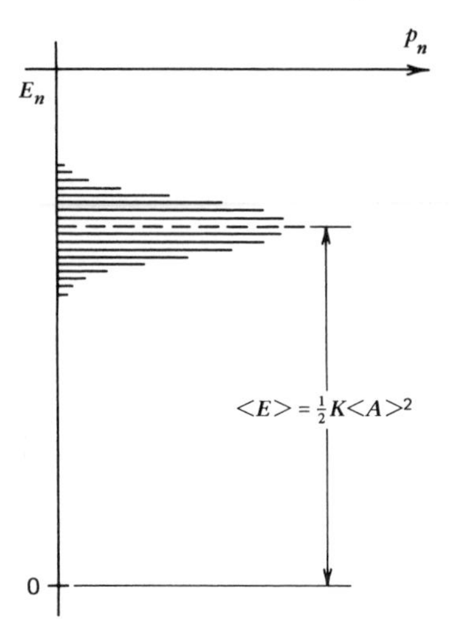

Figure 7.7. Eigenvalues of an oscillator as a function of their probability of occurrence.

What most likely will emerge is that the probability will have a maximum close to the average energy $\langle E \rangle$ of the classical oscillator. By performing a large set of these measurements for the oscillator, we could describe the experimental results by the function

$$p_n = \frac{\langle \mathbf{N} \rangle^n e^{-\langle \mathbf{N} \rangle}}{n!} \tag{7.47}$$

where

$$\langle \mathbf{N} \rangle = \frac{\langle E \rangle}{\hbar\omega} - \frac{1}{2} \tag{7.48}$$

and

$$n = \frac{E_n}{\hbar\omega} - \frac{1}{2} \tag{7.49}$$

Equation (7.47) is a Poisson distribution. The use of this probability function is derived or justified for the most natural state of the oscillator, since its use leads to the minimum uncertainty state for the oscillator. The probability distribution has been measured experimentally for photons.‡

‡Arecchi, S. T., *Phys. Rev. Lett.* **15**, 915 (1965).

With this probability function, it is an easy task to write the state of the oscillator. The coefficients of the states are related by

$$p_n = c_n^* c_n = |c_n|^2 \tag{7.50}$$

so that

$$|c_n| = \frac{\langle \mathbf{N} \rangle^{n/2} e^{-\langle \mathbf{N} \rangle /2}}{(n!)^{1/2}} \tag{7.51}$$

The phases of the complex numbers c_n may be set equal to zero, giving

$$c_n = \frac{\langle \mathbf{N} \rangle^{n/2} e^{-\langle \mathbf{N} \rangle /2}}{(n!)^{1/2}} \tag{7.52}$$

These coefficients establish the mathematical description of the most natural state of our classical oscillator as

$$|s_{\substack{\text{most}\\\text{natural}}}\rangle = \sum_n c_n |n\rangle = \sum_n \frac{\langle \mathbf{N} \rangle^{n/2} e^{-\langle \mathbf{N} \rangle /2}}{(n!)^{1/2}} |n\rangle \tag{7.53}$$

This describes an oscillator which has an energy near the classical average energy for most of the measurements but also has some chance of turning up with any of the other energy eigenvalues. The further the eigenvalue is from the average, the lower is its chance or probability of occurring.

This state has a sinusoidally oscillating expectation value for x and p of the oscillator. This same state is used later to describe the sinusoidally varying electric field of laser light. Since laser light, because of its well-defined phase is called *coherent light*, we shall refer to this state as the *coherent state*.

7.5 PROPERTIES OF THE COHERENT STATE

An interesting property arises for this particular oscillator state. The lowering operator acting upon it gives

$$\mathbf{a}|s\rangle = \mathbf{a} \sum_{n=0}^{\infty} \frac{\langle \mathbf{N} \rangle^{n/2} e^{-\langle \mathbf{N} \rangle /2}}{(n!)^{1/2}} |n\rangle \tag{7.54}$$

Using $\mathbf{a}|n\rangle = n^{1/2}|n-1\rangle$ we may write

$$\mathbf{a}|s\rangle = \sum_{n=1}^{\infty} \frac{\langle \mathbf{N}\rangle^{n/2} e^{-\langle \mathbf{N}\rangle/2}}{[(n-1)!]^{1/2}} |n-1\rangle \tag{7.55}$$

which by setting $m = n - 1$ may be rewritten

$$\mathbf{a}|s\rangle = \sum_{m=0}^{\infty} \frac{\langle \mathbf{N}\rangle^{(m+1)/2} e^{-\langle \mathbf{N}\rangle/2}}{(m!)^{1/2}} |m\rangle \tag{7.56}$$

or

$$\mathbf{a}|s\rangle = \langle \mathbf{N}\rangle^{1/2} \left[\sum_{m=0}^{\infty} \frac{\langle \mathbf{N}\rangle^{m/2} e^{-\langle \mathbf{N}\rangle/2}}{(m!)^{1/2}} |m\rangle \right]$$

Since m is only a dummy variable, the expression lying inside the bracket of Eq. (7.56) is none other than the state $|s\rangle$. Therefore, we have finally

$$\mathbf{a}|s\rangle = \langle \mathbf{N}\rangle^{1/2}|s\rangle \tag{7.57}$$

which is an eigenvalue equation, showing that the coherent state $|s\rangle$ is an eigenstate of the lowering operator **a**.

We shall simplify the notation by defining

$$\alpha = \langle \mathbf{N}\rangle^{1/2} \tag{7.58}$$

Then α, as an eigenvalue of the operator **a**, can also label the state, and the eigenvalue equation may be written more clearly

$$\mathbf{a}|\alpha\rangle = \alpha|\alpha\rangle \tag{7.59}$$

This is the quantum eigenvalue equation for the operator acting on the most natural state, or as we have renamed it, the coherent state.

To find the value of $\langle \mathbf{H}\rangle$, we will use our new eigenvalue equation. Forming the product of the eigenvalue equation and its adjoint

$$\langle\alpha|\mathbf{a}^\dagger = \langle\alpha|\alpha \tag{7.60}$$

we have

$$\langle\alpha|\mathbf{a}^\dagger\mathbf{a}|\alpha\rangle = \alpha^2\langle\alpha|\alpha\rangle \tag{7.61}$$

Multiplying both sides by $\hbar\omega$ and assuming $|\alpha\rangle$ to be normalized to unity, we find

$$\langle \mathbf{H} \rangle = \langle \alpha | \hbar\omega \mathbf{a}^\dagger \mathbf{a} | \alpha \rangle = \hbar\omega\alpha^2 \tag{7.62}$$

which is the average energy of the oscillator.

Since

$$\langle E \rangle = \langle \mathbf{H} \rangle = \tfrac{1}{2} K \langle A \rangle^2 = \hbar\omega\alpha^2 \tag{7.63}$$

the eigenvalue α is proportional to the average amplitude, which shows the intimate connection between the quantum and classical description of the oscillator. The number α takes on the significance of an amplitude, where the square of that amplitude is the average number of quanta of energy $\langle \mathbf{N} \rangle$.

7.6 A MORE ABSTRACT COMPLEX COHERENT STATE

Whereas α takes on the appearance of an amplitude which is a real, measurable characteristic of the system, we must allow α in general to be complex, since the operator $\mathbf{a}$ is *not* a Hermitian operator. With α as a complex number, we will look again at the properties of the coherent state $|\alpha\rangle$.

The expectation value $\langle \mathbf{N} \rangle$ for the coherent state reveals that

$$\langle \mathbf{N} \rangle = \langle \alpha | \mathbf{a}^\dagger \mathbf{a} | \alpha \rangle = \alpha^* \alpha = |\alpha|^2 \tag{7.64}$$

that is, the average value of the number operator is the absolute square of the complex number α. By introducing the expansion in number states $|n\rangle$ for the coherent state $|\alpha\rangle$, we have

$$\langle \alpha | \mathbf{a}^\dagger \mathbf{a} | \alpha \rangle = \sum_{m,n} c_m^* c_n \langle m | \mathbf{a}^\dagger \mathbf{a} | n \rangle \tag{7.65}$$

which by the eigenvalue equation and orthogonality properties for the $|n\rangle$ state is

$$\langle \mathbf{N} \rangle = \sum_n p_n n = |\alpha|^2 \tag{7.66}$$

where $p_n = c_n^* c_n$. This clearly shows $|\alpha|^2$ to be the average value of the expected distribution for the eigenvalue n (Fig. 7.8). We express the proba-

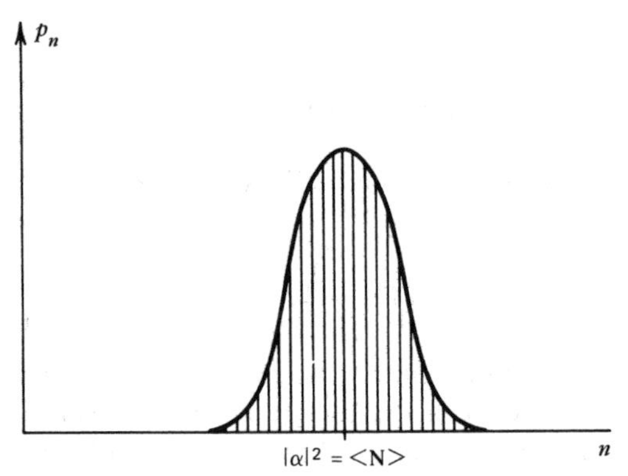

Figure 7.8. A Poisson distribution of the probability density for the number states of the oscillator.

bility of finding the eigenvalue in a single measurement for a complex α as

$$p_n = |c_n|^2 = \frac{e^{-|\alpha|^2}|\alpha|^{2n}}{n!} \tag{7.67}$$

which is the Poisson distribution again, but for a complex number.

Another interesting observation about this coherent state is seen if α is zero. The α state of lowest energy becomes

$$|\alpha = 0\rangle = |0\rangle \tag{7.68}$$

which is precisely the ground state of the oscillator we have already encountered, namely, the eigenstate of no quanta.

The complex number α appears as an eigenvalue in the eigenvalue equation

$$\mathbf{a}|\alpha\rangle = \alpha|\alpha\rangle \tag{7.69}$$

An eigenvalue associated with an observable operator, however, must be real. In this case, where it is complex, it is not a measurable quantity, nor is the operator with which it is associated an observable operator. Therefore $\mathbf{a}$ and $\mathbf{a}^\dagger$, although mathematically meaningful in terms of lowering and raising the quanta of energy, are not themselves observable or measurable in any experiment. We will return to this point again when quanta of energy are identified as the photons of an electromagnetic field. In the meantime α is a useful complex number whose absolute square is the average of the distribution of eigenvalues n for the α state.

REFERENCES

Anderson, E. E. *Modern Physics and Quantum Mechanics*, Saunders, Philadelphia, 1971.

Loudon, R. *The Quantum Theory of Light*, Oxford University Press, London, 1973.

Louisell, W. H. *Radiation and Noise in Quantum Electronics*, McGraw-Hill, New York, 1964.

Powell, J. L. and B. Crasemann. *Quantum Mechanics*, Addison-Wesley, Reading, Mass., 1961.

PROBLEMS

7.1. Evaluate the commutation relations $[\mathbf{n}, \mathbf{a}]$ and $[\mathbf{n}, \mathbf{a}^\dagger]$.

7.2. Show that the Hamiltonian of a harmonic oscillator can be written as $\mathbf{H} = \hbar\omega(\mathbf{a}\mathbf{a}^\dagger - \frac{1}{2})$.

7.3. From the eigenvalue equation $\mathbf{N}|n\rangle = n|n\rangle$, show the steps leading to $\mathbf{N}[\mathbf{a}|n\rangle] = (n-1)[\mathbf{a}|n\rangle]$.

7.4. Show that

$$\mathbf{a}^m|n\rangle = \left[\frac{n!}{(n-m)!}\right]^{1/2} |n-m\rangle$$

where n and m are integers.

7.5. Find the result of the operation $\mathbf{a}^\dagger|n-1\rangle$ where $n = 0$.

7.6. Prove that the raising operation is $\mathbf{a}^\dagger|n-1\rangle = \sqrt{n}\,|n\rangle$.

7.7. Verify the steps leading to Eq. (7.57).

7.8. In terms of the ground state, express the state of the oscillator as $|n\rangle = \sqrt{n!}\,\mathbf{a}^{\dagger n}|0\rangle$.

7.9. Prove that the coherent state can be written as $|\alpha\rangle = e^{(\alpha\mathbf{a}^\dagger - \frac{1}{2}|\alpha|^2)}|0\rangle$.

7.10. Show that the state $|s\rangle$ for $s = 0$ is the ground state.

7.11. Prove that $\langle\mathbf{N}\rangle$ is the same as $|\alpha|^2$.

8

Quantum Light

Now that we have the classical electromagnetic field energy in the same mathematical form as the energy of the classical harmonic oscillator, the quantization of the oscillator provides a direct scheme for an analogous quantization of the electromagnetic field. An immediate result of this process is the appearance of photons that describe the energy of the electromagnetic field, apparently resolving the dilemma of the wave–particle duality.

This success, however, comes at the expense of a simple but awkward set of quantum field energy states $|n\rangle$ that are difficult to relate to classical electromagnetic waves. These stationary solutions represent field states for which the number of photons is fixed and lead, through the uncertainty principle, to photons of totally unknown phase. Although historically this has been the starting point for quantum electrodynamics, the application of this straightforward quantization of field energy to problems in quantum optics, where the number of photons is usually large, has at best been tedious. Calculations, based on operators that change the number of photons by one, converge slowly making the application of perturbation techniques clumsy for optical fields near the classical limit.

An alternate theory tendered by Roy J. Glauber‡ leads to a more suitable set of states with which to describe the electromagnetic field and perform calculations relating to the coherence and correlation of photons. Glauber chooses to quantize the *field* rather than the energy, a process leading to states that describe the classically coherent wave of stable phase, namely, the wave describing ideal laser light. At the same time, these states not only predict the average number of photons for this field but also intimately relate the energy of the photons to the square of the classical field amplitudes.

These new states turn out to be the coherent oscillator states originally derived by Schrödinger, but long avoided as a basis for describing radiation since they are not orthogonal. As a complete set of field states, however, any

‡Glauber, R. J. *Phys. Rev.* **131**, 2766–2788 (1963).

arbitrary state of the field can be expressed in terms of coherent states, especially incoherent or random Gaussian light. Most importantly, with these states as a basis, a quantum definition of the correlation function emerges in a particularly convenient form which provides an entry for the analysis of coherence and interference for fields of photons and thereby, a fully quantum treatment of light.

8.1 A CLOSE LOOK AT WAVES AND OSCILLATORS

When we formulate the classical description of a plane traveling electromagnetic wave (Section 5-2), a single component of the field

$$\mathcal{E} = \mathcal{E}_0 \sin(kx - \omega t) \tag{8.1}$$

written in terms of the complex time-dependent amplitudes as

$$\mathcal{E} = i\left(\frac{\hbar\omega}{2\epsilon_0 L}\right)^{1/2}\left[a^*(t)e^{-ikx} - a(t)e^{ikx}\right] \tag{8.2}$$

renders the classical Hamiltonian of the electromagnetic field in the form

$$E = \hbar\omega a^* a \tag{8.3}$$

In retrospect, it is easier to see the motivation for this maneuver of redefinition, namely, to show the formal mathematical similarity between the variables describing the wave and those of the oscillator. Considering just one value (mode) of the field, we may display this analogy between the field and oscillator as follows.

Field	Oscillator
1. A plane traveling electromagnetic wave has an electric field $\mathcal{E} = \omega q(t)\sin kx + p(t)\cos kx$ where $p(t)$ and $q(t)$ are the time-dependent amplitudes having the form $p(t) = \dot{q}(t) = -\mathcal{E}_0 \sin \omega t$	1. The time-dependent momentum and position for the oscillator of unit mass, where $q(t) = q_0 \cos \omega t$ and $p_0 = q_0\omega$ are related by $p(t) = \dot{q}(t) = -p_0 \sin \omega t$

(8.4)

2. The energy of the field in terms of these functions, using $P = (\epsilon_0 L)^{1/2} p$ and $Q = (\epsilon_0 L)^{1/2} q$ is

$$E = \tfrac{1}{2}(P^2 + \omega^2 Q^2)$$

3. Defining complex amplitudes

$$a(t) = (2\hbar\omega)^{-1/2}(\omega Q + iP) = a(0)e^{-i\omega t}$$
$$a^*(t) = (2\hbar\omega)^{-1/2}(\omega Q - iP) = a(0)e^{i\omega t}$$

the positive and negative field parts are

$$\mathcal{E}^{(-)} = i\left(\frac{\hbar\omega}{2\epsilon_0 L}\right)^{1/2}[a^*(t)e^{-ikx}]$$

$$\mathcal{E}^{(+)} = -i\left(\frac{\hbar\omega}{2\epsilon_0 L}\right)^{1/2}[a(t)e^{ikx}]$$

4. The field energy becomes

$$E = 2\epsilon_0 \int_0^L \mathcal{E}^{(-)}\mathcal{E}^{(+)} dx = \hbar\omega a^* a$$

2. The classical oscillator Hamiltonian for this momentum and position is

$$H = \tfrac{1}{2}(p^2 + \omega^2 q^2) \quad (8.5)$$

3. Combining momentum and position to form the complex functions

$$a(t) = (2\hbar\omega)^{-1/2}(\omega q + ip)$$
$$a^*(t) = (2\hbar\omega)^{-1/2}(\omega q - ip) \quad (8.6)$$

we have

$$q = \left(\frac{\hbar}{2\omega}\right)^{1/2}[a(t) + a^*(t)]$$
$$p = -i\left(\frac{\hbar\omega}{2}\right)^{1/2}[a(t) + a^*(t)] \quad (8.7)$$

4. The oscillator energy becomes

$$H = \hbar\omega a^* a \quad (8.8)$$

This identification of the mathematical features provides the basis from which many of the mathematical quantum results for the oscillator may be applied to the field. While the application is not difficult, the interpretation must be done with care, since the field and oscillator are physically two different entities. In the case of the field, it is the amplitude which is oscillating sinusoidally with time at any given position x as expressed by

$$\mathcal{E}(t) = \mathcal{E}_0 \sin(\omega t + \theta) \quad (8.9a)$$

For the harmonic oscillator, it is the position of the mass which has the sinusoidal motion

$$q(t) = q_0 \cos(\omega t + \theta) \quad (8.9b)$$

Our task is to formulate a quantum picture of the electromagnetic field by giving the classical field equations a quantum interpretation from our understanding of the quantum oscillator.

The quantum analysis of the harmonic oscillator grew out of quantizing the energy through the energy eigenvalue postulate

$$\mathbf{H}|E_n\rangle = E_n|E_n\rangle \tag{8.10}$$

and the commutator property

$$\mathbf{qp} - \mathbf{pq} = i\hbar \tag{8.11}$$

solved most handily in terms of the raising and lowering operators which obey the commutation rule

$$\mathbf{aa}^\dagger - \mathbf{a}^\dagger\mathbf{a} = 1 \tag{8.12}$$

Using these relationships with the Hamiltonian of the oscillator expressed by

$$\mathbf{H} = \hbar\omega\left(\mathbf{a}^\dagger\mathbf{a} + \tfrac{1}{2}\right) = \hbar\omega\left(\mathbf{N} + \tfrac{1}{2}\right) \tag{8.13}$$

we were able to generate eigenvalues $E_n = \hbar\omega(n + \frac{1}{2})$ and the associated eigenstates

$$|n\rangle = \frac{\mathbf{a}^{\dagger n}}{\sqrt{n!}}|0\rangle \tag{8.14}$$

from the operator equation

$$\mathbf{N}|n\rangle = n|n\rangle \tag{8.15}$$

and the relations among the operators and states governed by

$$\mathbf{a}|0\rangle = 0 \tag{8.16a}$$

$$\mathbf{a}|n\rangle = \sqrt{n}\,|n-1\rangle \tag{8.16b}$$

$$\mathbf{a}^\dagger|n\rangle = \sqrt{n+1}\,|n+1\rangle \tag{8.16c}$$

Equipped with these oscillator results, we can proceed to the quantization of the electromagnetic field. By reinterpreting the expressions for the field amplitudes a and a^* as operator equations, complete with the properties of their oscillator counterparts, we can write the Hamiltonian operator for the kth mode of the *field* as

$$\mathbf{H}_k = \hbar\omega\left(\mathbf{a}_k^\dagger\mathbf{a}_k + \tfrac{1}{2}\right) = \hbar\omega\left(\mathbf{N}_k + \tfrac{1}{2}\right) \tag{8.17}$$

The eigenvalue equation for the field becomes

$$\mathbf{H}_k|E_{n_k}\rangle = E_{n_k}|E_{n_k}\rangle \tag{8.18}$$

where $|E_{n_k}\rangle$ represents an eigenstate of the field.

To solve Eq. (8.18) we use the same procedure as we used with the harmonic oscillator solution, except that for simplicity we redefine the zero of energy so that for the field, the eigenvalue is

$$n_k = \frac{E_{n_k}}{\hbar\omega_k} \tag{8.19}$$

and the number operator becomes

$$\mathbf{N}_k = \mathbf{a}_k^\dagger \mathbf{a}_k = \frac{\mathbf{H}_k}{\hbar\omega_k} \tag{8.20}$$

Thus, we obtain an equation familiar in form but new in meaning, namely

$$\mathbf{N}_k|n_k\rangle = \mathbf{a}_k^\dagger \mathbf{a}_k|n_k\rangle = n_k|n_k\rangle \tag{8.21}$$

where $|n_k\rangle$ is the nth eigenstate of the field in its kth mode. (In the following material, where only one mode of the field is considered, we drop the subscript k.) It now remains to interpret these results for the field.

8.2 THE MAKING OF A PHOTON

For the oscillator we have discussed the number of quanta of energy associated with each energy eigenvalue

$$E_n = \hbar\omega\left(n + \tfrac{1}{2}\right) \tag{8.22}$$

If we formalize $\hbar\omega$ as the basic quantum of energy, which for the electromagnetic field we call a photon, a quantum interpretation emerges.

We may look upon the field as a set of noninteracting photons. For instance, a field prepared in the state $|n\rangle$ has precisely n independent photons whose total energy (having redefined the zero of energy by the ground state value $\frac{1}{2}\hbar\omega$) is

$$E_n = n\hbar\omega \tag{8.23}$$

where n is the eigenvalue associated with a measurement of the number operator $\mathbf{N}$, thereby allowing any number of photons to occupy the same state. The raising or lowering operators, which act on an eigenstate of the field to change the state by one, are cast in the role of adding or removing a photon. As such they will be called the *creation and annihilation operators* in performing

$$\mathbf{a}^{\dagger}|n\rangle = \sqrt{n+1}\,|n+1\rangle \tag{8.24a}$$

and

$$\mathbf{a}|n\rangle = \sqrt{n}\,|n-1\rangle \tag{8.24b}$$

Consistent with this interpretation is the action of the annihilation operator

$$\mathbf{a}|0\rangle = 0 \tag{8.25}$$

where the vacuum state $|0\rangle$ represents a field of no photons, so that

$$|n\rangle = \frac{\mathbf{a}^{\dagger n}}{\sqrt{n!}}|0\rangle \tag{8.26}$$

can be looked upon as the creation operator acting n times upon the empty state of the field to produce the state of n photons.

The positive and negative parts of the classical field defined in terms of the amplitudes a and a^*, are now also associated with the operators $\mathbf{a}$ and $\mathbf{a}^{\dagger}$, thereby becoming the quantum field operators most conveniently expressed for a cavity of length L by

$$\mathcal{E}^{(-)} = i\left(\frac{\hbar\omega}{2}\right)^{1/2}\mathbf{a}^{\dagger}(t)e^{-ikx} \tag{8.27a}$$

and

$$\mathcal{E}^{(+)} = -i\left(\frac{\hbar\omega}{2}\right)^{1/2}\mathbf{a}(t)e^{ikx} \tag{8.27b}$$

The operator $\mathcal{E}^{(-)}$ creates photons of frequency ω and wave number k while the operator $\mathcal{E}^{(+)}$ removes photons from the field. Taken together as an ordered product of operators they yield the quantum field Hamiltonian operator

$$\mathbf{H} = 2\mathcal{E}^{(-)}\mathcal{E}^{(+)} = \hbar\omega\mathbf{a}^{\dagger}\mathbf{a} \tag{8.28}$$

the expectation value of which for a field in the eigenstate $|n\rangle$ is

$$\langle \mathbf{H} \rangle = \langle n | \hbar\omega \mathbf{a}^\dagger \mathbf{a} | n \rangle = n\hbar\omega \tag{8.29}$$

Thus, the average energy of the field equals the energy corresponding to an exact number of photons for that state.

This would complete our description of the eigenstates of the field in terms of photons and it would suffice if electromagnetic fields were always in eigenstates. But this is simply not the case. As easy as the field quantization has been up to this point, the nature of the electromagnetic field prepared in the eigenstate $|n\rangle$ is mysterious. We have quantized the energy of the field but not the field itself. The energy eigenstate is a stationary, time-independent state representing a fixed number of photons, or in other words, a fixed energy. Although we have defined operators for the field, we have not found an eigenvalue of the field operator (nor does one exist) for these particular states.

In order to understand the situation better, we invoke the uncertainty principle, but in a form pertaining to energy and time rather than momentum and position. This can be thought of in a number of ways. When we performed a transformation of the time-dependent variables $p(t)$ and $q(t)$ to the variables $a(t)$ and $a^*(t)$, the Hamiltonian took the operator form

$$\mathbf{H} \sim \mathbf{a}^\dagger(0)\mathbf{a}(0) e^{i\omega t} e^{-i\omega t} \tag{8.30}$$

Associating the number operator with the product $\mathbf{a}^\dagger(0)\mathbf{a}(0)$, and associating a phase operator $\boldsymbol{\phi}$ with ωt, we might expect the quantities Δn and $\Delta\phi$ to be related in an uncertainty relation. To find the nature of this relation, we must give consideration to the energy and time variables. Fourier transformations led to the relation

$$\Delta k \, \Delta x \simeq 1 \tag{8.31}$$

for packet widths in terms of wave number and spatial spread. Recalling the definition of group velocity as $v = \Delta\omega/\Delta k$, or in quantum terms $\Delta E/\Delta p$, and allowing that the center of a moving packet advances a distance $\Delta x = v\,\Delta t$, where v is the group velocity, we see that (8.31) can take the classical form

$$\Delta\omega \, \Delta t \simeq 1 \tag{8.32}$$

or the quantum form

$$\Delta E \, \Delta t \simeq \hbar \tag{8.33}$$

The same result can also be understood directly from the properties of Fourier transforms for spectral analysis and composition. In any case the quantum rendition of the spread relates an uncertainty in an observation of the energy with an uncertainty in the time for that observation. The interpretation for photons in a number state where $E = n\hbar\omega$ becomes

$$\Delta E = \hbar\omega\,\Delta n \tag{8.34a}$$

and

$$\Delta t = \frac{\Delta\phi}{\omega} \tag{8.34b}$$

that is, the energy uncertainty, is tied to the uncertainty in the number of photons present while the time for an observation relates to the spread in values of the phase describing the photon observation. Since the number of photons in the eigenstate $|n\rangle$ is an eigenvalue, the result of a measurement of the phase associated with an observation of the eigenvalue must, by the uncertainty relation, be completely undetermined.

This consequence leaves the condition of the field rather difficult to imagine, even in the classical limit of large numbers of photons. Since the phase is connected with time and the number of photons is certain, the field might take on a static appearance while traveling with constant amplitude and undeterminable phase (Fig. 8.1). Although mathematically this field can exist, it is not one that can be identified with properties of any known form of radiation.

The opposite extreme for a quantized wave presents an equally baffling picture in that a state of perfectly fixed phase $\Delta\phi = 0$, must be linked with

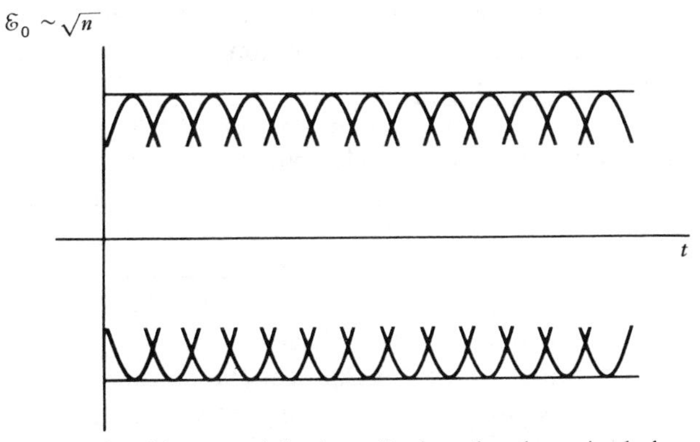

Figure 8.1. An unconscionable wave of fixed amplitude and undetermined phase associated with the state $|n\rangle$.

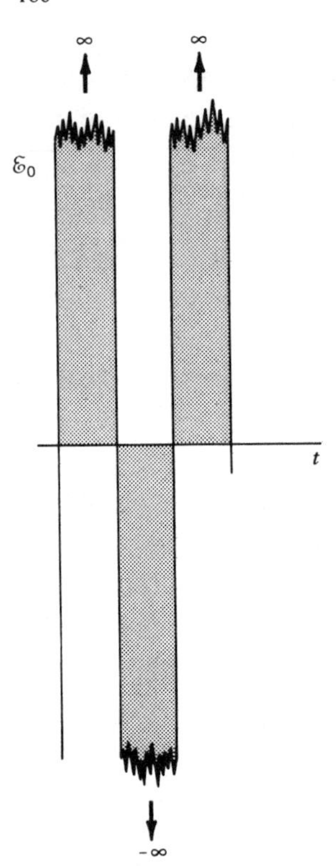

Figure 8.2. An equally unimaginable wave of precisely fixed phase and completely undetermined amplitude associated with a phase state.

total uncertainty in Δn, the number of photons present. Such a field would have in each half-cycle any energy between zero and infinity, fluctuating freely for times approaching zero (Fig. 8.2).

Thus, although we have defined operators and have found states for which photons are predicted as the quantum reality of an electromagnetic wave, it becomes necessary to seek states for which the field approaches a more familiar form, at least in the classical limit. In quantum terms, we wish to seek states for which the field itself is an eigenvalue of the field operator.

What we might expect, again following the oscillator example, is that a field can be in a most natural state, in particular, $|\alpha\rangle$, the coherent state. But while $|\alpha\rangle$ is the most natural state for an arbitrarily prepared oscillator, preparing the field in this state can be difficult. The field is composed of photons. Each individual photon is created in an independent atomic process in which there is little control over the phases of the emitted light. Only for laser light is phase control of the photons possible. There we find a light field which can be described as coherent, and, in fact, in the state $|\alpha\rangle$.

8.3 QUANTIZING THE ELECTROMAGNETIC FIELD

The classical electromagnetic field in a cavity of length L, written in terms of positive and negative frequency parts is

$$\mathcal{E}(x, t) = \mathcal{E}^{(-)} + \mathcal{E}^{(+)} = i\left(\frac{\hbar\omega}{2}\right)^{1/2}(a^* e^{-ikx} - ae^{ikx}) \tag{8.35}$$

and quantum cavity field operators which depend on the creation and annihilation operators are

$$\boldsymbol{\mathcal{E}}^{(-)} = i\left(\frac{\hbar\omega}{2}\right)^{1/2}\mathbf{a}^{\dagger} e^{-ikx} \tag{8.36a}$$

and

$$\boldsymbol{\mathcal{E}}^{(+)} = -i\left(\frac{\hbar\omega}{2}\right)^{1/2}\mathbf{a} e^{ikx} \tag{8.36b}$$

Since either part, $\mathcal{E}^{(-)}$ or $\mathcal{E}^{(+)}$, of the classical field $\mathcal{E}(x, t)$ is also a solution of Maxwell's equations, it is not unusual to expect that the operators (8.36) should have eigenvalues relating to some aspect of the classical field $\mathcal{E}(x, t)$, especially in the limit of large photon numbers. Thus, rather than quantize the field energy, which previously led to number states and eigenvalues of the energy operator, we will quantize the field in search of eigenvalues and states of the field operator.

By the quantum postulate we seek a solution for the field in the form

$$\boldsymbol{\mathcal{E}}|\mathcal{E}\rangle = \mathcal{E}|\mathcal{E}\rangle \tag{8.37a}$$

or the adjoint

$$\langle\mathcal{E}|\boldsymbol{\mathcal{E}}^{\dagger} = \langle\mathcal{E}|\mathcal{E}^* \tag{8.37b}$$

where the unknown state is labeled, as usual, by the eigenvalue of the operator. By choosing the field operator to be the operator of the positive frequency part $\boldsymbol{\mathcal{E}}^{(+)}$ we are led to the expression

$$\mathbf{a}|\alpha\rangle = \alpha|\alpha\rangle \tag{8.38}$$

where the state is again labeled by the eigenvalue α, which in this case is the replacement for

$$\alpha = \frac{\mathcal{E}^{(+)}(x, t)}{-i\left(\frac{\hbar\omega}{2}\right)^{1/2} e^{ikx}} \tag{8.39}$$

making the eigenvalue of the field operator

$$\mathcal{E}^{(+)}(x, t) = -i\left(\frac{\hbar\omega}{2}\right)^{1/2} \alpha e^{ikx} \tag{8.40}$$

Similarly, the adjoint field equation (8.37b), where $(\mathcal{E}^{(+)})^\dagger = \mathcal{E}^{(-)}$, leads to the eigenvalue

$$\mathcal{E}^{(-)}(x, t) = i\left(\frac{\hbar\omega}{2}\right)^{1/2} \alpha^* e^{-ikx} \tag{8.41}$$

through the adjoint relation

$$\langle\alpha|\mathbf{a}^\dagger = \langle\alpha|\alpha^* \tag{8.42}$$

Thus the eigenvalue equations for the field produce familiar counterparts. Equations (8.38) and (8.42) have already described the coherent state of the oscillator where α made its appearance through a unique combination of number states leading to a most probable or least uncertain state of the oscillator. The eigenvalue α in this application was also related through its absolute square to the average or again, most probable energy of the oscillator prepared in the state $|\alpha\rangle$.

As a linear superposition of number states, the coherent state of the oscillator was given by

$$|\alpha\rangle = \sum_n c_n|n\rangle = \sum_n \frac{|\alpha|^n e^{-(|\alpha|^2)/2}}{\sqrt{n!}}|n\rangle \tag{8.43}$$

where the probability of occurrence for measuring a particular energy eigenvalue of the oscillator number states is expressed by the square of the coefficients c_n and, in particular, is

$$p_n = |c_n|^2 = \frac{|\alpha|^{2n} e^{-|\alpha|^2}}{n!} \tag{8.44}$$

The eigenvalue α is unusual in that it may be a complex number which for the oscillator is of little concern since the amplitude operators **a** and **a**† are not directly connected to any classically observable property. Finally, for the oscillator, the absolute square of the eigenvalue is the average energy after many measurements as expressed by

$$\langle E \rangle = \hbar\omega|\alpha|^2 \tag{8.45}$$

A translation of these properties to the quantized electromagnetic field yields an interesting insight into the characteristics of the states $|\alpha\rangle$ for the field, gives a role to the eigenvalue α that resolves the wave–photon paradox, and opens the door to an analysis of optical fields prepared in any state.

8.4 COHERENT STATE OF THE FIELD

The quantization of the field expressed by

$$\mathcal{E}^{(+)}|\alpha\rangle = \mathcal{E}^{(+)}|\alpha\rangle \tag{8.46a}$$

or the adjoint

$$\langle\alpha|\mathcal{E}^{(-)} = \langle\alpha|\mathcal{E}^{(-)} \tag{8.46b}$$

produces the field eigenvalues $\mathcal{E}^{(+)}$ and $\mathcal{E}^{(-)}$ in terms of α and α^*, where α takes on the role of a complex field amplitude. The sum of the eigenvalues, however, is real and in fact becomes

$$\mathcal{E} = \mathcal{E}^{(+)} + \mathcal{E}^{(-)} = -i\left(\frac{\hbar\omega}{2}\right)^{1/2}\left(\alpha e^{ikx} - \alpha^* e^{-ikx}\right) \tag{8.47}$$

which upon expressing the time-dependent amplitude α as

$$\alpha = |\alpha| e^{-i\omega t} \tag{8.48}$$

shows the sum of the eigenvalues (8.47) to be the real classical cavity field

$$\mathcal{E} = (2\hbar\omega)^{1/2}|\alpha| \sin(kx - \omega t) \tag{8.49}$$

or more explicitly

$$\mathcal{E} = (\epsilon_0 L)^{1/2}\mathcal{E}_0 \sin(kx - \omega t) \tag{8.50}$$

Thus the stable classical field has an amplitude related to an eigenvalue of the quantum state $|\alpha\rangle$. This is the coherent wave of controlled phase capable of being produced by an ideal laser. For this reason the quantum field state $|\alpha\rangle$ is called the coherent or pure state for light.

The product of the eigenvalue equations for the coherent state is

$$\langle\alpha|\mathcal{E}^{(-)}\mathcal{E}^{(+)}|\alpha\rangle = \mathcal{E}^{(-)}\mathcal{E}^{(+)}\langle\alpha|\alpha\rangle \tag{8.51}$$

which for the normalized state, namely, $\langle\alpha|\alpha\rangle = 1$, is the expectation value of the product of the field operators. But this product $\mathcal{E}^{(-)}\mathcal{E}^{(+)}$ as in Eq. (8.28), is simply related to the Hamiltonian for the field by

$$\mathbf{H} = 2\mathcal{E}^{(-)}\mathcal{E}^{(+)} = \hbar\omega\mathbf{a}^{\dagger}\mathbf{a} \tag{8.52}$$

allowing the expectation value of the energy to be written as

$$\langle\mathbf{H}\rangle = 2\langle\alpha|\mathcal{E}^{(-)}\mathcal{E}^{(+)}|\alpha\rangle = 2\mathcal{E}^{(-)}\mathcal{E}^{(+)} \tag{8.53}$$

With the field eigenvalues, Eqs. (8.40) and (8.41), the expectation value or average energy of the field is

$$\langle\mathbf{H}\rangle = 2\left[i\left(\frac{\hbar\omega}{2}\right)^{1/2}\alpha^{*}e^{-ikx}\right]\left[-i\left(\frac{\hbar\omega}{2}\right)^{1/2}\alpha e^{ikx}\right] = \hbar\omega|\alpha|^{2} \tag{8.54}$$

Since the expectation value of the number operator is

$$\frac{\langle\mathbf{H}\rangle}{\hbar\omega} = \langle\mathbf{N}\rangle \tag{8.55}$$

we have $|\alpha|^2 = \langle\mathbf{N}\rangle$, which now is interpreted as the average number of photons in the coherent field. Thus, we have arrived at a fully quantum picture of a coherent radiation field as a collection of photons whose average number after many measurements is proportional to the average energy of the field.

The result (8.54) finally puts to rest the photon paradox, showing not only that light is composed of photons but that in the limit of large numbers of photons, where the fluctuation ΔN is small compared to $\langle\mathbf{N}\rangle$, the wave representing the coherent photon field is the classically stable wave (Fig. 8.3). In fact, the coherent state $|\alpha\rangle$ is a minimum uncertainty state for which the product of the fluctuations in number and phase is a minimum. As such, it is the best compromise between the two previously discussed cases, namely, the state of precise number and indeterminable phase, and the opposite extreme, precise phase and indefinite number. However, this

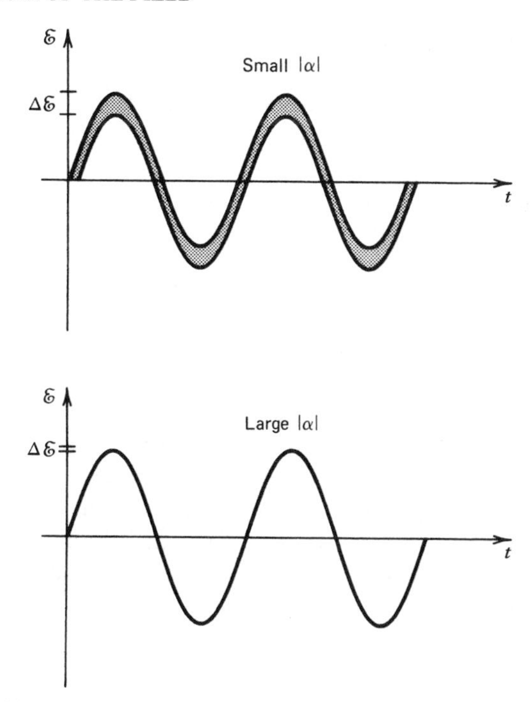

Figure 8.3. Shrinking the vertical scale for large $|\alpha|$ makes the relative uncertainty less than for the wave with small $|\alpha|$.

translation of the uncertainty properties is not to be taken too literally. On one hand, it does show the coherent state $|\alpha\rangle$ to be the only quantum state to correspond to the classically coherent sinusoidal electromagnetic wave. But even when the number of photons is small, the photons will still exhibit perfect coherence as long as they are prepared in the quantum coherent state $|\alpha\rangle$. In detection processes where practically single photons are counted, there is no corresponding wave. However, each photon capable of interfering with itself retains its coherence properties.

A measurement of intensity is related to the average energy $\langle \mathbf{H} \rangle$ through the time Δt allotted to detecting $|\alpha|^2$ worth of photons crossing perpendicularly an instrument with an eye of area A. A photodetector with a short response time is capable of counting photons in terms of the photoelectrons produced to give a measure of the photon energy flow. Using the classical definition for intensity as

$$I = \frac{1}{A} \frac{\Delta E}{\Delta t} \tag{8.56}$$

where $\Delta E = \langle \mathbf{H} \rangle$, we have for coherent light

$$I = \frac{1}{A} \frac{\hbar\omega\langle \mathbf{N} \rangle}{\Delta t} = \frac{1}{A} \frac{\hbar\omega|\alpha|^2}{\Delta t} \tag{8.57}$$

If for convenience we set the area and time interval to unity the intensity is given by

$$I = \hbar\omega|\alpha|^2 = \hbar\omega\langle \mathbf{N} \rangle \tag{8.58}$$

Thus we have come back full circle to one of the original paradoxical observations from the photoelectric experiment which gave birth to the idea of the photon in the first place. But now there is no paradox. Using the notion of a classical wave and its energy and the quantum mechanical oscillator as a guide to constructing quantum field operators, we have derived, at least for coherent fields, the nonparadoxical quantum observations that the intensity of a wave increases with the absolute square of the amplitude α, or equivalently, the intensity is directly proportional to the average number of photons in the observed field.

8.5 PURE STATES, PACKETS, AND MINIMUM UNCERTAINTY

Until now we have taken on faith that the coherent or pure state is equivalent to the most probable state of the oscillator. It remains not only to prove this identity but also to show that the state $|\alpha\rangle$ is a minimum uncertainty state. Starting with the quantization condition for the field

$$\mathbf{a}|\alpha\rangle = \alpha|\alpha\rangle \tag{8.59}$$

and recognizing the expansion in terms of the complete set of number states $|n\rangle$ as

$$|\alpha\rangle = \sum_n c_n|n\rangle \tag{8.60}$$

we have, taking the scalar product of (8.60) with a number state

$$\langle n'|\alpha\rangle = \sum_n c_n\langle n'|n\rangle = c_{n'} \tag{8.61}$$

since all the number states are orthogonal and normalized, that is, $\langle n'|n\rangle =$

$\delta_{n', n}$. This enables (8.60) to be written as

$$|\alpha\rangle = \sum_n |n\rangle\langle n|\alpha\rangle \tag{8.62}$$

where $\langle n|\alpha\rangle = c_n$. Evaluating the coefficient by forming the product of Eq. (8.59) with $\langle n|$, we have

$$\langle n|\mathbf{a}|\alpha\rangle = \alpha\langle \alpha|\alpha\rangle \tag{8.63}$$

With the aid of the raising operation

$$\mathbf{a}^\dagger|n\rangle = (n+1)^{1/2}|n+1\rangle \tag{8.64a}$$

but expressed in its adjoint form

$$\langle n|\mathbf{a} = (n+1)^{1/2}\langle n+1| \tag{8.64b}$$

Eq. (8.63) becomes

$$c_n = \langle n|\alpha\rangle = \frac{(n+1)^{1/2}}{\alpha}\langle n+1|\alpha\rangle \tag{8.65}$$

Since the number state has already been expressed by Eq. (7.36) as a raised version of the ground state, namely,

$$|n\rangle = \frac{\mathbf{a}^{\dagger n}}{(n!)^{1/2}}|0\rangle \tag{8.66a}$$

the adjoint of this expression written in terms of the number $n+1$ is

$$\langle n+1| = \langle 0|\frac{\mathbf{a}^{n+1}}{[(n+1)!]^{1/2}} \tag{8.66b}$$

Substituting (8.66b) into Eq. (8.65) we have

$$c_n = \frac{(n+1)^{1/2}}{\alpha}\langle 0|\frac{\mathbf{a}^{n+1}}{[(n+1)!]^{1/2}}|\alpha\rangle \tag{8.67}$$

which simplifies to

$$c_n = \langle 0|\alpha\rangle\frac{\alpha^n}{(n!)^{1/2}} = c_0\frac{\alpha^n}{(n!)^{1/2}} \tag{8.68}$$

Thus, the expansion for the coherent state is

$$|\alpha\rangle = \sum_n c_n|n\rangle = \sum_n c_0 \frac{\alpha^n}{(n!)^{1/2}}|n\rangle \tag{8.69}$$

Normalizing the state $|\alpha\rangle$, recalling that the n states are orthogonal, and that both c_0 and α may be complex numbers, leads to

$$1 = \langle\alpha|\alpha\rangle = |c_0|^2 \sum_n \sum_{n'} \frac{\alpha^n}{(n!)^{1/2}} \frac{\alpha^{*n'}}{(n'!)^{1/2}} \langle n'|n\rangle \tag{8.70a}$$

$$= |c_0|^2 \sum_n \frac{|\alpha|^{2n}}{n!} \tag{8.70b}$$

$$= |c_0|^2 e^{|\alpha|^2} \tag{8.70c}$$

where in the last step, the sum is evaluated by the series $\sum_n x^n/n! = e^x$ with $x = |\alpha|^2$. Fixing the phase of either c_0 or $|\alpha\rangle$ arbitrarily to zero since they share the same phase yields

$$c_0 = e^{-|\alpha|^2/2} \tag{8.71}$$

and the coherent state from Eq. (8.69) as

$$|\alpha\rangle = e^{-|\alpha|^2/2} \sum_n \frac{\alpha^n}{(n!)^{1/2}}|n\rangle \tag{8.72}$$

showing the most probable state of the oscillator to indeed be the coherent state of the electromagnetic field for which the probability of finding the average number of photons in the nth state is shaped by the Poisson distribution

$$|c_n|^2 = |\langle n|\alpha\rangle|^2 = \frac{|\alpha|^{2n}}{n!} e^{-|\alpha|^2} = \frac{\langle \mathbf{N}\rangle e^{-\langle \mathbf{N}\rangle}}{n!} \tag{8.73}$$

It now remains to show that the coherent state is a minimum uncertainty state, thereby uniquely allowing it to correspond as closely as possible to the classically stable wave in the limit of large $|\alpha|^2$. To demonstrate this case we return to the coherent quantum state condition

$$\mathbf{a}|\alpha\rangle = \alpha|\alpha\rangle \tag{8.74a}$$

or

$$(2\hbar\omega)^{-1/2}(\omega\mathbf{q} + i\mathbf{p})|\alpha\rangle = \alpha|\alpha\rangle \tag{8.74b}$$

where the annihilation operator **a** has been expressed in terms of its dependence on the operators **p** and **q**. With the appearance of the operator **q**, which for the oscillator is the coordinate position operator but for the field has no direct physical counterpart, we must introduce another quantizing condition

$$\langle q|\mathbf{q} = \langle q|q \tag{8.75}$$

which defines the action of the operator **q** on the eigenstate $\langle q|$ where q is the eigenvalue of the coordinate operator. Recalling that the momentum operator **p** (again, there is no physical counterpart for the field) is defined by $-i\hbar\partial/\partial q$, we can form

$$\langle q|\mathbf{p} = -i\hbar\frac{\partial}{\partial q}\langle q| \tag{8.76}$$

as an aid to obtaining the product of $\langle q|$ with Eq. (8.74b), namely,

$$(2\hbar\omega)^{-1/2}\left(\omega\mathbf{q} + \hbar\frac{\partial}{\partial q}\right)\langle q|\alpha\rangle = \alpha\langle q|\alpha\rangle \tag{8.77a}$$

which is the differential equation

$$\frac{\partial}{\partial q}\langle q|\alpha\rangle = -2\left(\frac{\omega}{2\hbar}\right)^{1/2}\left[\left(\frac{\omega}{2\hbar}\right)^{1/2}q - \alpha\right]\langle q|\alpha\rangle \tag{8.77b}$$

At this point it is convenient to introduce the *coherent wave function*

$$\psi_\alpha(q) \equiv \langle q|\alpha\rangle \tag{8.78}$$

which can be visualized as the projection of the state vector $|\alpha\rangle$ onto an infinite continuous set of orthogonal coordinate state vector components. This makes $\psi_\alpha(q)$ a function representing the product of those projections for a continuous range of values of the spatial coordinate q. Inserting the wave function in the differential equation (8.77b), we can solve for it by performing the integration

$$\int\frac{d\psi_\alpha}{\psi_\alpha} = -2\int u\,du \tag{8.79}$$

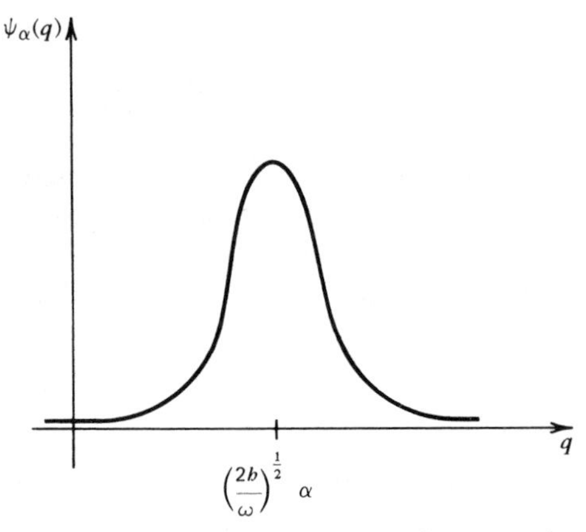

Figure 8.4. Time-independent Gaussian-shape wave function.

where $u = (\omega/2\hbar)^{1/2}q - \alpha$. This yields a Gaussian-shaped wave function

$$\psi_\alpha(q) = A\exp\left\{-\left[\left(\frac{\omega}{2\hbar}\right)^{1/2} q - \alpha\right]^2\right\} \tag{8.80}$$

Upon application of the normalization condition

$$\int_{-\infty}^{\infty} \psi_\alpha^* \psi_\alpha \, dq = 1 \tag{8.81}$$

the time-independent wave function becomes (Fig. 8.4)

$$\psi_\alpha(q) = \left(\frac{\omega}{\pi\hbar}\right)^{1/4} \exp\left\{-\left[\left(\frac{\omega}{2\hbar}\right)^{1/2} q - \alpha\right]^2\right\} \tag{8.82}$$

For this time-independent result the value of α is also taken at $t = 0$, which makes α the same as either $|\alpha|$ or $\mathrm{Re}\,\alpha$. A similar derivation using $\langle p|\mathbf{q} = i\hbar(\partial/\partial p)\langle p|$ produces the time-independent momentum wave packet

$$\phi_\alpha(p) = (\pi\hbar\omega)^{-1/4} \exp\left\{-\left[(2\hbar\omega)^{-1/2} p + i\alpha\right]^2\right\} \tag{8.83}$$

which is also the Fourier transform of $\psi_\alpha(q)$.

Using explicitly the time dependence of the operator $\mathbf{a}(t) = \mathbf{a}e^{-i\omega t}$, from the way in which it was originally defined, allows the time-dependent eigenvalue to be expressed by $\alpha(t) = \alpha e^{-i\omega t}$ making each eigenstate expressed as $|\alpha e^{-i\omega t}\rangle$.[‡] If we replace α by $\alpha e^{-i\omega t}$ in Eq. (8.82) and we take the absolute square of the wave functions, the time-dependent probability densities become

$$\psi^*(q,t)\psi(q,t) = \left(\frac{\omega}{\pi\hbar}\right)^{1/2}\exp\left\{-2\left[\left(\frac{\omega}{2\hbar}\right)^{1/2}q - \alpha\cos\omega t\right]^2\right\} \tag{8.84a}$$

and

$$\phi^*(p,t)\phi(p,t) = (\pi\hbar\omega)^{-1/2}\exp\left\{-2\left[(2\hbar\omega)^{-1/2}p + \alpha\sin\omega t\right]^2\right\} \tag{8.84b}$$

The centers of these functions oscillate $\pi/2$ out of phase with one another in simple harmonic motion with a frequency ω and amplitudes $(2\hbar/\omega)^{1/2}\alpha$ for the coordinate packet and $(2\hbar\omega)^{1/2}\alpha$ for the momentum packet (Fig. 8.5).[§]

To show that the coherent state is one of minimum uncertainty, we must find the width of these Gaussian functions by evaluating the root mean square definitions

$$\Delta q = \langle(\mathbf{q} - \langle\mathbf{q}\rangle)^2\rangle^{1/2} = (\langle\mathbf{q}^2\rangle - \langle\mathbf{q}\rangle^2)^{1/2} \tag{8.85a}$$

and

$$\Delta p = \langle(\mathbf{p} - \langle\mathbf{p}\rangle)^2\rangle^{1/2} = (\langle\mathbf{p}^2\rangle - \langle\mathbf{p}\rangle^2)^{1/2} \tag{8.85b}$$

Using the definition of the operators as

$$\mathbf{q} = \left(\frac{\hbar}{2\omega}\right)^{1/2}(\mathbf{a}^\dagger + \mathbf{a}) \tag{8.86a}$$

and

$$\mathbf{p} = -i\left(\frac{\hbar\omega}{2}\right)^{1/2}(\mathbf{a}^\dagger - \mathbf{a}) \tag{8.86b}$$

and showing the expectation values of the creation and annihilation opera-

[‡]Schrodinger, E. *Die Naturwissenshaften* **14**, 664 (1926).
[§]Goldin, E. and J. Bregman. *Swinging Quanta*, International Film Bureau, Inc., Chicago, 1982.

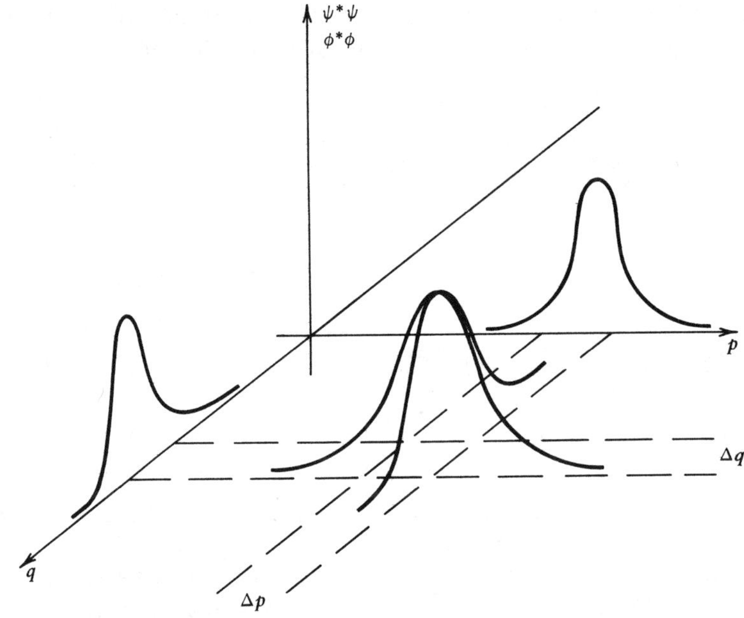

Figure 8.5. The probability densities in phase space.

tors to be

$$\langle\alpha|\mathbf{a}|\alpha\rangle = \langle\mathbf{a}\rangle = \langle\mathbf{a}^\dagger\rangle^* = \alpha e^{-i\omega t} \tag{8.87a}$$

$$\langle\alpha|\mathbf{a}^2|\alpha\rangle = \langle\mathbf{a}^2\rangle = \langle\mathbf{a}^{\dagger 2}\rangle^* = \alpha^2 e^{-2i\omega t} \tag{8.87b}$$

$$\langle\alpha|\mathbf{a}^\dagger\mathbf{a}|\alpha\rangle = \langle\mathbf{a}^\dagger\mathbf{a}\rangle = \langle\mathbf{a}\mathbf{a}^\dagger\rangle - 1 = \alpha^2 \tag{8.87c}$$

we see that the average values of the position and momentum operators become

$$\langle\mathbf{q}\rangle = \alpha\left(\frac{2\hbar}{\omega}\right)^{1/2}\cos\omega t \tag{8.88a}$$

$$\langle\mathbf{p}\rangle = \alpha(2\hbar\omega)^{1/2}\sin\omega t \tag{8.88b}$$

and

$$\langle\mathbf{q}^2\rangle = \langle\mathbf{q}\rangle^2 + \frac{\hbar}{2\omega} \tag{8.89a}$$

$$\langle\mathbf{p}^2\rangle = \langle\mathbf{p}\rangle^2 + \frac{\hbar\omega}{2} \tag{8.89b}$$

Finally, an application of these results to the uncertainty principle gives

$$\Delta q\,\Delta p = \hbar/2 \tag{8.90}$$

which is the minimum uncertainty possible, showing not only that the coherent state exhibits the best localization simultaneously for both momentum and coordinate position but also that there is no spreading of the wave packets in time.

8.6 COHERENCE, CORRELATION, AND QUANTUM OPTICS

For a cavity field in the pure state, the energy was found to be

$$2\langle\alpha|\mathcal{E}^{(-)}\mathcal{E}^{(+)}|\alpha\rangle = \hbar\omega\langle\mathbf{N}\rangle \tag{8.91}$$

By performing a similar calculation but with field operators for two different space–time points

$$\mathcal{E}_1^{(-)}(\mathbf{r}_1, t_1) = i\left(\frac{\hbar\omega}{2}\right)^{1/2} e^{-i\mathbf{k}\cdot\mathbf{r}_1}\mathbf{a}^\dagger(t_1) \tag{8.92a}$$

and

$$\mathcal{E}_2^{(+)}(\mathbf{r}_2, t_2) = -i\left(\frac{\hbar\omega}{2}\right)^{1/2} e^{i\mathbf{k}\cdot\mathbf{r}_2}\mathbf{a}(t_2) \tag{8.92b}$$

we obtain the quantity

$$\langle\alpha|\mathcal{E}_1^{(-)}\mathcal{E}_2^{(+)}|\alpha\rangle = \frac{\hbar\omega}{2}\langle\mathbf{N}\rangle e^{i\mathbf{k}\cdot(\mathbf{r}_2-\mathbf{r}_1)}e^{i\omega(t_2-t_1)} \tag{8.93}$$

which is in the form of the interference term we are accustomed to from Young's interference experiment. Having previously defined the classical first-order correlation function for this type of calculation, we are now in a position to render a quantum definition of first-order coherence for one mode of the pure state of the field as

$$G_{12}^{(1)} = \langle\alpha|\mathcal{E}_1^{(-)}\mathcal{E}_2^{(+)}|\alpha\rangle = \mathcal{E}_1^{(-)}\mathcal{E}_2^{(+)} \tag{8.94}$$

This definition demonstrates the advantage offered by the use of the coherent states and positive and negative operators. Occurring in pairs, the field operators act back and forth on the state $|\alpha\rangle$, allowing a convenient definition of correlation functions of all orders of coherence as

$$G^{(n)}(\mathbf{r}_1 t_1, \ldots, \mathbf{r}_n t_n; \mathbf{r}_{n+1} t_{n+1}, \ldots, \mathbf{r}_{2n} t_{2n}) = \langle \alpha | \mathcal{E}_1^{(-)} \cdots \mathcal{E}_n^{(-)} \mathcal{E}_{n+1}^{(+)} \cdots \mathcal{E}_{2n}^{(+)} | \alpha \rangle \tag{8.95}$$

No longer restricting the field to a single mode, we can construct a Fourier series for any multimode light field as

$$\mathcal{E}^{(+)}(\mathbf{r}, t) = -i \sum_{\mathbf{k}} \left(\tfrac{1}{2}\hbar\omega\right)^{1/2} e^{i\mathbf{k}\cdot\mathbf{r}} \mathbf{a}_{\mathbf{k}}(t) \tag{8.96}$$

Since the photons of each mode are noninteracting, the state for the multimode field can be written as the product of states for each mode, namely,

$$|\alpha\rangle = |\alpha_{\mathbf{k}_1}\rangle |\alpha_{\mathbf{k}_2}\rangle \cdots \tag{8.97a}$$

for which we introduce a notation for the product

$$|\alpha\rangle = \prod_{\mathbf{k}} |\alpha_{\mathbf{k}}\rangle = |\{\alpha_{\mathbf{k}}\}\rangle \tag{8.97b}$$

The operators $\mathbf{a}_{\mathbf{k}}$ and $\mathbf{a}_{\mathbf{k}}^{\dagger}$ for each component of the Fourier series pick out the appropriate single mode coherent state to act upon, making the first-order correlation function for the multimode field become

$$G^{(1)} = \langle \{\alpha_{\mathbf{k}}\} | \mathcal{E}^{(-)} \mathcal{E}^{(+)} | \{\alpha_{\mathbf{k}}\} \rangle \tag{8.98}$$

As an example of the second-order coherence of the type encountered in the Hanbury Brown and Twiss experiment for light fields from two space–time points, the correlation function is expressed by

$$\begin{aligned} G_{12}^{(2)} &= \langle \alpha | \mathcal{E}^{(-)}(\mathbf{r}_1, t_1) \mathcal{E}^{(-)}(\mathbf{r}_2, t_2) \mathcal{E}^{(+)}(\mathbf{r}_2, t_2) \mathcal{E}^{(+)}(\mathbf{r}_1, t_1) | \alpha \rangle \\ &= \mathcal{E}_1^{(-)} \mathcal{E}_2^{(-)} \mathcal{E}_2^{(+)} \mathcal{E}_1^{(+)} \end{aligned} \tag{8.99}$$

Thus the definition of the correlation function of any light field which is in the state $|\{a_{\mathbf{k}}\}\rangle$ (such as the pure state of light as encountered in an ideal laser beam) causes the correlation functions to factorize into

$$G^{(n)} = \prod_{i=1}^{n} \mathcal{E}_i^{(-)} \cdot \prod_{j=n+1}^{2n} \mathcal{E}_j^{(+)} \tag{8.100}$$

allowing the pure state of light to be fully coherent to all orders of the correlation function.

If a light field is not in a pure state but rather some state $|s\rangle$, the correlation function becomes

$$G^{(n)} = \langle s|\mathbf{O}_n|s\rangle \tag{8.101}$$

where $\mathbf{O}_n$ stands for the ordered pairs of field operators appropriate to the order of the correlation. When the coherent states $|\alpha\rangle$ are used as a basis and the general state $|s\rangle$ is expanded in terms of the basis set of states, the correlation function for fields of large photon numbers becomes

$$G^{(n)} = \sum_{\alpha} p(\alpha)\langle\alpha|\mathbf{O}_n|\alpha\rangle \tag{8.102}$$

which is an ensemble average of the operator for a statistical mixture of pure states. This can be done since any state can be expanded in terms of a basis set if that set is complete. The coherent set of states actually form an overcomplete set. That they are not orthogonal is tolerable since widely separated states are approximately orthogonal, and as long as $p(\alpha)$ is slowly varying which would be the case near the classical limit for chaotic random light.

In the case of thermal light, therefore, which is the most common light encountered, the possibility of finding a particular state $|\alpha\rangle$ among the mixed states $|s\rangle$ is given by the function $p(\alpha)$, which is taken to be the Gaussian distribution

$$p(\alpha) = \frac{1}{\pi\langle\mathbf{N}\rangle} e^{-|\alpha|^2/\langle\mathbf{N}\rangle} \tag{8.103}$$

Of particular interest, because it relates to many first-order optical interference experiments, is the form of the correlation function for a single mode of the Gaussian mixed states, namely,

$$G_{12}^{(1)} = \int_0^\infty p(\alpha)\{(\tfrac{1}{2}\hbar\omega)e^{i\mathbf{k}\cdot(\mathbf{r}_2-\mathbf{r}_1)}e^{-i\omega(t_2-t_1)}\}|\alpha|^2\,d\alpha \tag{8.104}$$

With the aid of the integral

$$\int_0^\infty e^{-|\alpha|^2}|\alpha|^{2n}\,d\alpha = \pi n! \tag{8.105}$$

which is an integration performed in the complex plane, the first-order correlation function becomes

$$G_{12}^{(1)} = \frac{\hbar\omega}{2}\langle\mathbf{N}\rangle e^{i\mathbf{k}\cdot(\mathbf{r}_2-\mathbf{r}_1)}e^{-i\omega(t_2-t_1)} \tag{8.106}$$

showing thereby that Gaussian light is first-order coherent. But that is all that it can be. The second-order correlation function does not factor, and in fact, for two space–time points takes a form which is the sum of the products of the first-order functions

$$\begin{aligned} G_{12}^{(2)} &= \left[\mathcal{E}_1^{(-)}\mathcal{E}_2^{(+)}\right]\cdot\left[\mathcal{E}_2^{(-)}\mathcal{E}_1^{(+)}\right] + \left[\mathcal{E}_1^{(-)}\mathcal{E}_1^{(+)}\right]\cdot\left[\mathcal{E}_2^{(-)}\mathcal{E}_2^{(+)}\right] \\ &= \left[G_{12}^{(1)}\right]^2 + \left[G_{11}^{(1)}\right]^2 \end{aligned} \tag{8.107}$$

Thus Gaussian light cannot be second-order coherent. The Hanbury Brown and Twiss experiment actually measures the deviation from second-order coherence for random light. A derivation of any higher order correlation functions for Gaussian light results in $G^{(n)}$ expressed as a sum of products of first-order correlation functions.

So while Gaussian light can show the common interference effects now familiar from Young's experiment, it cannot exhibit higher order coherence. No matter how well filtered or collimated, chaotic light consists of photons governed by quantum statistics to behave always as random Gaussian light. Laser light, on the other hand, (which is the only other known state in which light can be prepared) is fundamentally different. It consists of photons capable of demonstrating high orders of coherence represented by a statistical function

$$p(\alpha) = \delta(\alpha) \tag{8.108}$$

differing dramatically from the probability function for Gaussian light.

We have reached the entrance point for correlation calculations and the definition of coherence in its fundamental quantum garb, which can be expressed in general for any kind of light field as

$$G^{(n)} = \sum_{\alpha} p(\{\alpha_{\mathbf{k}}\})\langle\{\alpha_{\mathbf{k}}\}|\mathbf{O}_n|\{\alpha_{\mathbf{k}}\}\rangle \tag{8.109}$$

This is the beginning for quantum optical analysis of experiments which produce, count, and correlate photons, and holds the possibility of representing new types of light fields and their effects when or if they are conceived in the future.

This introduction to quantum optics also beckons one to study such topics as the production of laser light and photodetection counting devices. Given the newness of this field, its connection to the general field of quantum electrodynamics, and future application in optical physics, continuation of study in this area is well worth the effort.

REFERENCES

Glauber, R. J. *Phys. Rev.* **131**, 2766–2788 (1963).

Loudon, R. *The Quantum Theory of Light*, Oxford University Press, London, 1973.

Loisell, W. H. *Radiation and Noise in Quantum Electronics*, McGraw–Hill, New York, 1964.

Panofsky, W. K. H. and M. Phillips. *Classical Electricity and Magnetism*, Addison-Wesley, Reading, Mass., 1955.

PROBLEMS

8.1. Evaluate the classical field energy for a cavity in the form

$$E = 2\epsilon_0 \int_0^L \mathcal{E}^{(-)}\mathcal{E}^{(+)}\, dx$$

8.2. Two different coherent states having the eigenvalues α and α' are only approximately orthogonal. Derive the relation

$$|\langle \alpha | \alpha' \rangle|^2 = e^{-|\alpha - \alpha'|^2}$$

which expresses this condition.

8.3. Derive the result indicated by the normalization procedure $\langle \alpha | \alpha \rangle = |c_0|^2 e^{|\alpha|^2}$.

8.4. Obtain the expression for the coherent state

$$|\alpha\rangle = e^{-|\alpha|^2/2} \sum_n \frac{\alpha^n}{(n!)^{1/2}} |n\rangle$$

Can this result be written in terms of the ground state $|0\rangle$? Explain.

8.5. Derive the time-independent momentum wave packet (8.33).

8.6. Derive the Gaussian wave function for the oscillator $\psi_\alpha(q)$, as in Eq. (8.82).

8.7. Show the explicit steps in arriving at the relations (8.89a) and (8.89b).

8.8. Derive the minimum uncertainty result for the coherent (Gaussian) wave functions, namely, $\Delta q\, \Delta p = \hbar/2$.

8.9. For Gaussian light, if all the space–time coordinates are the same, show that

$$G^{(n)}(r, t, \ldots, r, t; r, t, \ldots, r, t) = n!\left[G^{(1)}(r, t; r, t)\right]^n$$

8.10. Evaluate the integral over the entire complex plane, showing

$$\int_0^\infty e^{-|\alpha|^2}|\alpha|^{2n}\,d\alpha = \pi n!$$

where the element of area is

$$d\alpha = d|\alpha| \cdot |\alpha|\,d\theta$$

by first showing that it can be written as

$$\int_0^\infty |\alpha|^{2n+1}e^{-|\alpha|^2}d|\alpha| \int_0^{2\pi} d\theta$$

8.11. Using the expansions for the coherent state, show that

$$\int |\alpha\rangle\langle\alpha|\,d\alpha = \pi\sum_n |n\rangle\langle n|$$

8.12. With the result of the previous problem, obtain

$$\frac{1}{\pi}\int |\alpha\rangle\langle\alpha|\,d\alpha = 1$$

8.13. Explain the result in Problem 8.12 in terms of the completeness theorem. Can an arbitrary state of the system be expanded in terms of the coherent state? Give an example.

Index